Trace-Element Geochemistry of Coal Resource Development Related to Environmental Quality and Health

PANEL ON THE TRACE ELEMENT GEOCHEMISTRY OF COAL RESOURCE DEVELOPMENT RELATED TO HEALTH
Subcommittee on the Geochemical Environment in Relation to Health and Disease
U.S. National Committee for Geochemistry

Assembly of Mathematical and Physical Sciences
National Research Council

NATIONAL ACADEMY PRESS
Washington, D.C. 1980

NOTICE: The project that is the subject of this report was approved by the Governing Board of the National Research Council, whose members are drawn from the Councils of the National Academy of Sciences, the National Academy of Engineering, and the Institute of Medicine. The members of the Committee responsible for the report were chosen for their special competences and with regard for appropriate balance.

This report has been reviewed by a group other than the authors according to procedures approved by a Report Review Committee consisting of members of the National Academy of Sciences, the National Academy of Engineering, and the Institute of Medicine.

The National Research Council was established by the National Academy of Sciences in 1916 to associate the broad community of science and technology with the Academy's purposes of furthering knowledge and of advising the federal government. The Council operates in accordance with general policies determined by the Academy under the authority of its Congressional charter of 1863, which establishes the Academy as a private, nonprofit, self-governing membership corporation. The Council has become the principal operating agency of both the Academy of Sciences and the National Academy of Engineering in the conduct of their services to the government, the public, and the scientific and engineering communities. It is administered jointly by both Academies and the Institute of Medicine. The Academy of Engineering and the Institute of Medicine were established in 1964 and 1970, respectively, under the charter of the Academy of Sciences.

Library of Congress Cataloging in Publication Data

U.S. National Committee for Geochemistry. Panel on the Trace Element Geochemistry of Coal Resource Development Related to Health.
Trace-element geochemistry of coal resource development related to environmental quality and health.

Bibliography: p.
Includes index.
1. Coal--Environmental aspects. 2. Coal--Toxicology. 3. Trace elements--Environmental aspects. I. Title.
TD195.C58U55 1980 628.5 80-15305
ISBN 0-309-03048-X

Available from

National Academy Press
2101 Constitution Avenue, N.W.
Washington, D.C. 20418

Printed in the United States of America

PANEL ON THE TRACE ELEMENT GEOCHEMISTRY OF COAL RESOURCE DEVELOPMENT RELATED TO HEALTH (PECH)

Bobby G. Wixson, Civil Engineering Department, Environmental Research Center, University of Missouri, Rolla, Chairman
Albert L. Page, Kearney Foundation of Soil Science and Department of Soil and Environmental Sciences, University of California, Riverside, Vice Chairman
Jack L. Beckner, Mining Consultant, Environmental Programs, Denver, Colorado
Leonard D. Hamilton, Medical Research Center, Brookhaven National Laboratory, Upton, New York
Richard C. Neavel, Synthetic Fuels, Exxon Research and Engineering Co., Baytown, Texas
Rodney R. Ruch, Analytical Chemistry Section, Illinois State Geological Survey, Urbana

Liaison Members

Pope A. Lawrence, Office of Federal Activities, Environmental Protection Agency, Washington, D.C.
Vernon E. Swanson, Geologic Division, U.S. Geological Survey, Denver, and Geological Society of America, Boulder, Colorado

Special Consultants

M. E. Baser, Brookhaven National Laboratory, Upton, New York
A. A. Elseewi, Department of Soil and Environmental Sciences, University of California, Riverside
S. C. Morris, Brookhaven National Laboratory, Upton, New York
I. R. Straughan, Southern California Edison Company, Rosemead, California

Staff

William L. Petrie, Executive Secretary
Fred Block, Staff Officer
June R. Galke, Secretary

SUBCOMMITTEE ON THE GEOCHEMICAL ENVIRONMENT IN RELATION TO HEALTH AND DISEASE (GERHD)

Leonard T. Kurland, Department of Medical Statistics, Epidemiology, and Genetics, Mayo Clinic, Rochester, Minnesota, Co-Chairman

Albert L. Page, Kearney Foundation of Soil Science and Department of Soil and Environmental Sciences, University of California, Riverside, Co-Chairman

Harold J. Gluskoter, Department of Coal Research, Exxon Production Research Company, Houston, Texas

Rolf Hartung, Department of Environmental and Industrial Health, University of Michigan, Ann Arbor

Donald J. Horvath, Division of Animal and Veterinary Sciences, West Virginia University, Morgantown

Everett A. Jenne, Water Resources Division, U.S. Geological Survey, Menlo Park, California

Walter D. Keller, Department of Geology, University of Missouri, Columbia

Ruth Patrick, Department of Limnology, Academy of Natural Sciences, Philadelphia, Pennsylvania

Harold G. Petering, Kettering Laboratory, University of Cincinnati, Cincinnati, Ohio

Walter J. Pories, Department of Surgery, East Carolina University, Greenville, North Carolina

Ex Officio

Rosemary J. Vidale, Chairman, U.S. National Committee for Geochemistry, Los Alamos Scientific Laboratory, Los Alamos, New Mexico

U.S. NATIONAL COMMITTEE FOR GEOCHEMISTRY

Rosemary J. Vidale, Los Alamos Scientific Laboratory, Los Alamos, New Mexico, Chairman
Robert A. Berner, Yale University, New Haven, Connecticut
John R. Castaño, Shell Development Company, Houston, Texas
Maria Luisa B. Crawford, Bryn Mawr College, Bryn Mawr, Pennsylvania
Harold J. Gluskoter, Exxon Production Research Company, Houston, Texas
V. Rama Murthy, University of Minnesota, Minneapolis
James L. Papike, State University of New York, Stony Brook
Arthur W. Rose, Pennsylvania State University, University Park
William M. Sackett, University of South Florida, St. Petersburg
E-an Zen, U.S. Geological Survey, Reston, Virginia

Ex Officio

George W. Wetherill, President, International Association of Geochemistry and Cosmochemistry; Carnegie Institution of Washington, Washington, D.C.

PREFACE

This report assesses for decision makers and those involved in coal resource development the environmental and health impacts of trace-element effects arising from significant increases in the use of coal, unless unusual precautions are invoked.

Increasing demands for energy and the pressing need for decreased dependence of the United States on imported oil require greater use of coal to meet the nation's energy needs during the next decade. If coal production and consumption are increased at a greatly accelerated rate, concern arises over the release, mobilization, transportation, distribution, and assimilation of certain trace elements, with possible adverse effects on the environment and human health. Although most trace elements are beneficial and many are essential, the Panel is concerned with those that may become available to the environment in toxic amounts. It is, therefore, important to understand their geochemical pathways from coal and rocks via air, water, and soil to plants, animals, and ultimately humans, and their relation to health and disease.

To address this problem, the Subcommittee on the Geochemical Environment in Relation to Health and Disease (GERHD) of the U.S. National Committee for Geochemistry, Assembly of Mathematical and Physical Sciences, National Research Council (NRC), established the Panel on Trace Element Geochemistry of Coal Resource Development Related to Health (PECH) with the following objectives:

- To determine the present state of knowledge of the composition of coal and potential health hazards of trace elements during development of coal resources;
- To examine the environmental and man-induced factors that are related to human health through the deficiencies or excesses of trace elements;
- To evaluate how much of an additional burden of natural or man-made pollution would be tolerable from both detrimental and beneficial trace elements;
- To disseminate information useful for the establishment of baseline studies and the development of guidelines or standards necessary for the elimination or control of toxic trace elements associated with coal development;

• To point out gaps in our knowledge about health effects of trace elements released by coal development and suggest what further studies or research might be needed to correct these deficiencies.

The Panel was formed with knowledgeable biomedical, environmental, and earth scientists and engineers from agencies, industries, and universities concerned with coal, with emphasis on trace-element environmental and health aspects.

Planning and conduct of this study were coordinated with other NRC committees undertaking related studies. The study on the Redistribution of Accessory Elements from the Increased Utilization of Coal by the NRC Committee on Accessory Elements (National Academy of Sciences, Washington, D.C., 1979) is in two volumes; Volume I is concerned chiefly with resources and economic aspects of coal and oil shale development. The Committee on Research Needs on the Health Effects of Fossil Fuel Combustion Products (HEFF) of the NRC Assembly of Life Sciences is chiefly concerned with the effects of fossil-fuel combustion products on human health. These three studies, which address somewhat different objectives, complement and supplement each other.

Certain assumptions were made by the Panel to highlight the central issues of trace elements and health and to avoid unwarranted duplication of other studies. These assumptions, which focus on specific areas of concern, were incorporated into the design of the report to help the users to understand the Panel's rationale. The assumptions are as follows:

• Coal production will increase in the United States at an accelerated rate, with the major portion being used for the generation of electricity.

• Most of the increased coal production will be from surface-mine operations in low-sulfur western coals.

• Current mining technology will gradually decrease the significance of acid mine drainage (AMD) from old mine operations in eastern and interior coals, so that potential problems with trace elements in these contexts should diminish rather than increase.

• Solid wastes or residues generated by increased coal production, coal-washing operations, and coal use will elevate the amounts of certain soluble trace elements entering the environment and pose a major disposal or control problem.

• Beneficial, as well as detrimental, trace-element occurrences must be evaluated geochemically, geographically, and geologically at site-specific locations with respect to both environmental conditions and human health effects.

• The conclusions and recommendations of this report should be of direct value in minimizing detrimental impacts on human health that may be expected from increases in the development of coal resources.

Based on the charge to the Panel and these assumptions, this report describes the amounts and distribution of trace elements related to the coal source; the various methods of coal extraction, preparation, transportation, and use; and the disposal or recycling of the remaining residues or wastes. The known or projected health effects are discussed at the end of each section.

The Panel's summary, conclusions, and recommendations are presented initially along with a ranked listing of the elements in order of estimated importance to this study. Recommendations are made concerning site-specific trace-element occurrences that need further study and concerning those trace elements that do not appear to merit additional consideration.

The following chapter of this report on trace elements deals with the formation of coal, its physical and chemical composition, the resources of coal, and its utilization. A portion of this chapter also deals with chemical analytical procedures and standards, with emphasis on possible future modifications and improvements. General health considerations are also presented.

The next chapter concerns coal mining, including the different surface and subsurface methods employed, subsidence and water control, soil and overburden handling and replacement, refuse and waste disposal, vegetation, and the health effects associated with the various coal-mining steps.

Chapter 4 deals with the cleaning, storage, and transportation of coal and the attendant occupational and public health effects.

Coal combustion products are discussed next in regard to the disposal of harmful trace elements and the health aspects concerned with their physical properties, their chemical composition, and their deposition from the atmosphere and disposal on soils or surface waters. The information on fly-ash-amended soils and plant growth emphasizes the concern for proper disposal of coal-waste residues.

The last chapter of the report deals with synthetic-fuel processes, the trace elements so released and/or concentrated, and the health effects associated with coking and synfuel processes.

The report concludes with references, supplemental material in appendixes, and an index.

Although important concerns have been expressed about increases of CO_2 in the atmosphere, acid precipitation, and radioactivity associated with additional coal development, these are not the topic of this report and consequently were not evaluated in detail.

ACKNOWLEDGMENTS

This report of the Panel on Trace Element Geochemistry of Coal Resource Development Related to Health (PECH) has been made possible by support from the U.S. Bureau of Mines and the U.S. Geological Survey under Contract S0144074, the National Cancer Institute under Contract NO1-CP-45616, the U.S. Atomic Energy Commission and the Energy Research and Development Administration (now the Department of Energy) through transfer of funds under National Science Foundation Contract NSF-C310 Task Order 206 and ERDA/DOE Contract EY76C02 2708 015, and Environmental Protection Agency Contract 68 03 2554, which greatly assisted with the completion of this report, its publication, and the ongoing work of the Subcommittee on the Geochemical Environment in Relation to Health and Disease. The financial support provided by these agencies is gratefully acknowledged.

Special thanks is extended to the many helpful organizations and individuals that provided information for and assisted with this report. Their efforts have been truly outstanding and very much appreciated. In addition to the universities, companies, or other organizations employing the members of the Panel, notable contributions have been made by Ian Straughan of the Southern California Edison Company, A. A. Elseewi of the University of California at Riverside, and M. E. Baser and S. C. Morris of Brookhaven National Laboratory, who served as special consultants.

The Subcommittee and the Panel acknowledge with special appreciation the efforts of the following prepublication reviewers: Mary O. Amdur, Massachusetts Institute of Technology, Cambridge, Massachusetts; John C. Crowell, University of California, Santa Barbara; W. Gary Ernst, University of California, Los Angeles; Harold J. Gluskoter, Exxon Production Research Company, Houston, Texas; Paul E. Morrow, University of Rochester, Rochester, New York; Ralph M. Perhac, Electric Power Research Institute, Palo Alto, California; and Harry Perry, Resources for the Future, Washington, D.C.

The Subcommittee and Panel also thank the members and staff of the other National Research Council committees whose cooperation and coordination helped to facilitate this report.

CONTENTS

SUMMARY, CONCLUSIONS, AND RECOMMENDATIONS

SUMMARY AND CONCLUSIONS

The U.S. coal resources are estimated at approximately 3.6 trillion metric tons, of which an estimated 396 billion metric tons can be mined economically. Annual consumption of coal in the United States is currently about 600 million metric tons, of which about two thirds are used in electric power generation. The production and use of coal in the United States is projected to double in the next decade. Electric power utilities are the major users of coal, and most of the projected increase will also be for the generation of power. Minor increases in coal use as a chemical feedstock to replace petroleum sources is anticipated, but not in the next few years. The use of coal in coking will probably remain at about the present level.

Because of the way it was formed, coal may contain every naturally occurring element. Except for the major elements (carbon, hydrogen, oxygen, nitrogen, sulfur, silicon, iron, aluminum, potassium, calcium, and magnesium), most elements are present in coal in trace amounts. The chemical composition of coal is site- and source-specific and varies with rank and location in a coal bed, as well as between beds. Of the trace elements in coal, only boron, mercury, lead, molybdenum, arsenic, cadmium, antimony, and selenium are commonly more abundant in coal than in the average composition of the earth's crust. However, not all are concentrated in coals from the various coal-mining regions of the United States.

Generally, coals from the western United States contain lower concentrations of sulfur and certain other trace elements than do coals from other regions in the country. In the future, more of the high- and medium-sulfur coals (>1 percent) will undergo cleaning and processing before use, primarily to reduce sulfur and inorganic matter concentrations. In the process of coal cleaning, trace-element concentrations are usually lower in comparison with those in uncleaned coals. Because western coals are generally low in sulfur (<1 percent), they are not expected to require coal cleaning on a broad scale.

The handling of coal during mining, cleaning, storage, and transportation produces materials that may release trace elements with possible health effects. Information on these trace-element releases is sparse. Coal cleaning generates some dust problems, but most critical is the proper control of slurries or residues produced during the washing and preparation processes. The slurries are acidic and have elevated trace-element concentrations, especially of arsenic, lead, zinc, manganese, gallium, and selenium.

Coal dust is a major problem encountered in storage and transport of coal. Aqueous solutions of surface-active agents are often used to minimize windblown dust. Soils near coal storage areas and along heavily used coal-haulage routes can, however, contain elevated trace-element concentrations because of additions of windblown coal dust or because of the leaching and dispersal action of rainfall.

Direct health impacts of dust or leachates from coal during cleaning, storage, or transportation are relatively unknown, but dust inhalation could present a problem if not controlled. Dust-control measures currently used in the United States are expected to prevent severe health hazards.

Residues remaining from coal cleaning and preparation using current methods are site-specific problems dependent on local geochemical conditions, climatic conditions, and the type of coal treated. Careful monitoring, neutralization, and disposal should effectively control potential health problems.

Unless new and currently unknown health-related problems emerge, present regulations and controls for coal cleaning, storage, or transportation with existing technology appear to be adequate to accommodate increased coal consumption.

Combustion of coal in power-generating plants produces a variety of residues and emissions (bottom ash, boiler slag, fly ash, flue-gas desulfurization sludge, and noncaptured gases and particulates). The relative amount of each residue depends on the power-plant configuration and emission-control devices used. Trace elements originating from the source coal are redistributed among the various residue streams. For example, the concentrations of most inorganic trace elements are higher in ash residues than in the coal. Elements that are volatile at combustion temperatures (e.g., selenium, molybdenum, mercury, boron) may or may not exhibit higher concentrations in fly ash than in coal.

Fly-ash residues captured in bag filters or electrostatic precipitators, which currently represent the greater percentage of solids collected, contain most of the trace-element burden. Research has demonstrated that when fly ash is incorporated in soil it is beneficial to crop growth because it serves as a source of some elements essential to plants, such as sulfur, calcium, molybdenum, boron, zinc, and possibly manganese. Fly ash is substantially alkaline in nature, and its potential as an inexpensive amendment to neutralize acidic soils has been demonstrated. Field and greenhouse studies, however, have shown possible detrimental effects on plant growth and crop quality where fly ash has been added to soils at rates greater than approximately 4 percent by weight. In particular, concentrations of total dissolved salts and boron may reach levels that are toxic to plants. Forage crops grown on fly-ash-amended soils may accumulate levels of molybdenum and selenium considered unsafe for livestock consumption. Data available indicate that hazardous accumulations of certain additional trace elements (such as cadmium, arsenic, lead, and mercury) by plants grown on fly-ash-amended soils will not occur.

In Figure 1, trace elements that are of concern in coal and residues produced from coal resource development have been placed in five categories: of greatest concern; of moderate concern; of minor concern; radioactive elements generally considered to be of minor concern but adequate information is lacking for proper assessment; and of concern but with negligible concentrations in coal and coal residues.

• Besides the troublesome gaseous compounds of carbon, nitrogen, and sulfur, the elements of greatest concern are arsenic, boron, cadmium, lead, mercury, molybdenum, and selenium. These eight elements commonly occur in coal and in residues from coal cleaning and combustion at concentrations greater than those encountered on the average in the earth's crust. The elements carbon, nitrogen, and sulfur and their compounds, although considered to be of greatest concern, are beyond the scope of this report and are not discussed here. Arsenic, cadmium, mercury, and lead are highly toxic to most biological systems when they occur in available form at concentrations above certain critical levels. Levels of molybdenum and selenium derived from airborne deposits resulting from coal development produce concentrations in soil and water that should present no adverse impact on the health of humans or plants. Even if molybdenum and selenium become somewhat enriched in soils, they

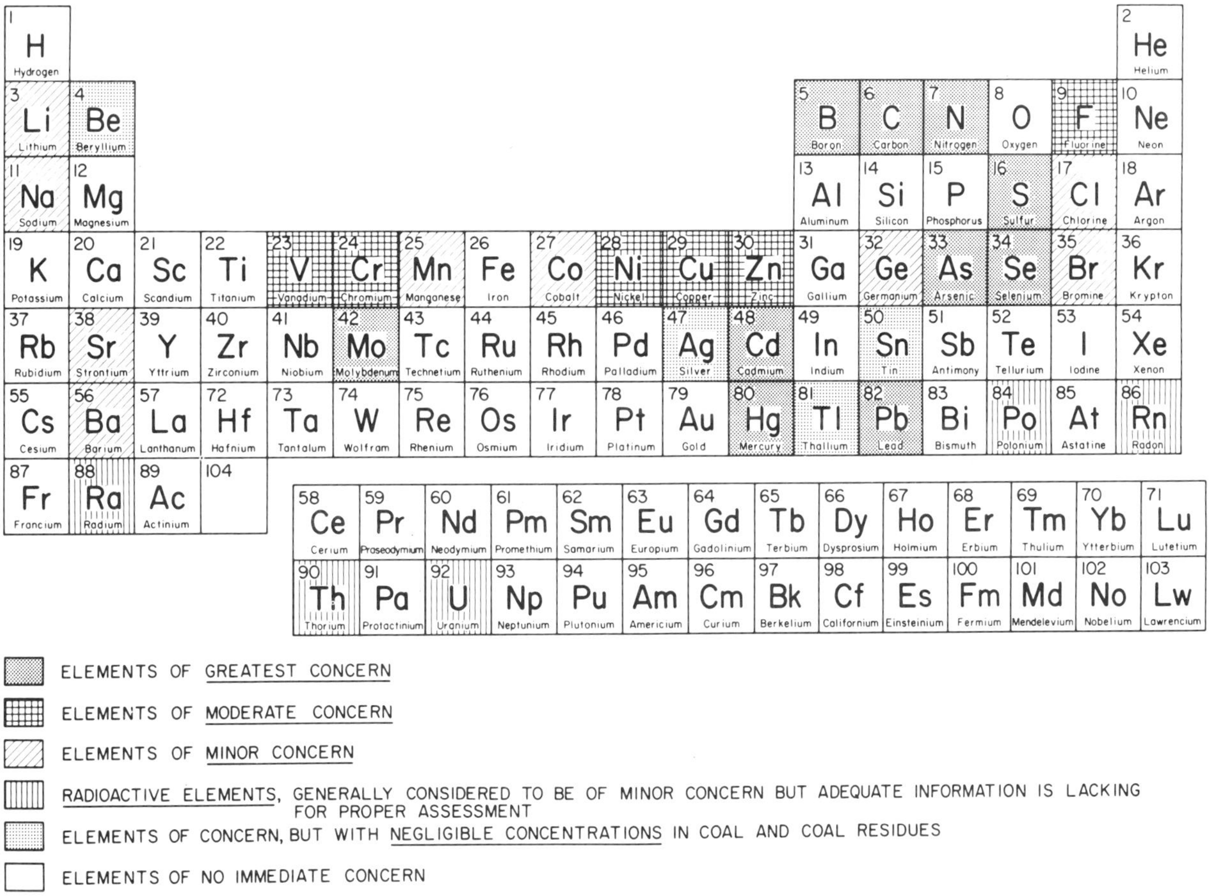

FIGURE 1 Potentially hazardous elements in coal resource development.

should not be toxic to plants. However, the disposal of fly ash on soils used to produce forage crops could lead to accumulations of these elements in the forage and render it unsafe for consumption by animals. Adverse impacts of excess boron contamination are limited to possible phytotoxic effects. However, atmospheric deposition of boron onto plants or soil is small, and the probability of encountering levels sufficiently elevated to cause phytotoxicity is remote.

• Elements considered by the Panel to be of moderate concern include chromium, vanadium, copper, zinc, nickel, and fluorine. The Panel's rationale for grouping these elements in this category is that they are potentially toxic and occur at elevated concentrations in some residues produced during coal combustion, but the probability of adverse impact is considered less than for the elements of greatest concern. Also, vegetation and animal health problems that could occur from excesses of these elements are reversible and can be corrected economically. Copper, nickel, and zinc, in certain site-specific situations, could accumulate over the years in soils to levels that are phytotoxic. Fluorine is a highly volatile element, and, in areas where its deposition onto vegetation is great, the forage may not be suitable for consumption by animals.

• Those elements considered to be of minor concern include barium, strontium, sodium, manganese, cobalt, antimony, lithium, chlorine, and bromine. Although these elements frequently occur in fly ash at concentrations greater than in the normal geochemical environment, the probability of a specific adverse impact that could not be easily corrected is regarded as remote.

• The radioactive elements ^{238}U, ^{235}U, and ^{232}Th are present in coal in small quantities and are released into the environment as airborne effluents, in collected residues [fly ash, bottom ash, and flue-gas desulfurization (FGD) sludge], and in mine tailings during the course of coal utilization. Because levels of radioactivity associated with these elements during coal resource development are not significantly different from those commonly found in nature, and because they are in a geochemically immobile form, their effects on health are thought to be negligible. Available information on other radioactive elements such as radon, radium, and polonium suggests that the effects of their radioactivity associated with coal resource development are minimal; however, because adequate information is lacking, the Panel has classed these radioactive elements in a category requiring additional information for conclusive assessment.

• Although there are other elements in the periodic chart with known and documented relationships to health, many of them have only negligible concentrations in coal and are commonly rendered inert in coal residues; therefore, potential hazards associated with coal development from these elements are considered to be remote. Elements in this category include tin, beryllium, thallium, silver, and tellurium. Environmental exposure resulting from coal resource development of the other remaining elements, not otherwise cited, is not generally regarded as hazardous to health.

Residues from the coal combustion process include those materials collected by the emission-control devices and those that escape these devices and enter the atmosphere. The noncaptured particulates entering the atmosphere from stacks are of submicrometer size. High percentages may remain airborne for extended periods of time; similarly, some elements (e.g., fluorine and mercury) are released in a gaseous form from stacks and remain airborne for some time. Current best-available technology is such that more than 99 percent of the solid residues from coal combustion are collected (Figure 2). Therefore, in the future the trace-element burden associated with coal resource development in terms of total quantities is one of proper disposal or recycling of the collected residues (Figures 3 and 4). New power plants should represent only a negligible contribution to the trace-element burden from airborne sources. However, power plants not equipped with modern control devices will continue to emit to the atmosphere considerable quantities of trace elements until they are phased out or required to install up-to-date equipment. The trace-element enrichment in soils and vegetation around these plants will continue to be localized and will occur in regions close to the source, generally within a few kilometers.

Current annual production of fly ash from the combustion of coal by electric utilities is 56 million tons, of which less than 15 percent is used in industrial operations. As relatively small amounts of the fly ash produced are used, and the amounts generated will increase as coal use increases, fly ash will accumulate and intensify disposal problems. Approximately 36 percent of the fly-ash disposal is by trucking to storage areas, presumably landfills, and the remainder is sluiced to disposal ponds. Leachate from unsealed disposal ponds could contaminate groundwater. Likewise, dust from dry ash storage areas, if left unstabilized, could contaminate soils in adjacent areas. Volunteer growth of native vegetation also may accumulate certain trace elements (molybdenum and selenium) to levels unsafe for consumption by wildlife. Current information on the potential problems associated with the disposal of fly ash is inadequate. Knowledge required to assess the impacts of fly-ash disposal, therefore, should be developed. Once developed, criteria for the safe disposal of fly ash should be formulated.

The submicrometer-sized particles emitted to the atmosphere from power plants (even though equipped with modern control devices) include those that, when inhaled, may be deposited in the lung. Concern has therefore been expressed over possible long-term, subacute, chronic poisoning from certain trace elements in the vicinity of coal-burning plants. The current data show that the contribution of particulates from a power plant to the trace-element concentration in air surrounding such plants is small, much less than other sources of natural and man-made atmospheric contamination, and because of particle size considerations inhalation of these submicrometer-sized particles should not be ignored.

Near-term development of coal in the United States will be principally at western sources where surface mining is widely practiced. In the process of topsoil removal, overburden removal, coal extraction, overburden and topsoil replacement, and regrading, the porosity of the displaced material is increased. This increased porosity may allow additional leaching of trace elements by groundwater or surface water. Because, except for boron, the solubilities of the trace elements of greatest concern in the neutral to alkaline rocks and soils of the western region are small, the transport of trace elements via this

FIGURE 2 The 2700-MW, coal-fired Bruce Mansfield Plant at Shippingport, Pennsylvania, is the world's largest electric generating station with comprehensive environmental controls. (Photo courtesy Pennsylvania Power Company.)

pathway should be limited to short distances. On a localized basis, certain specific imbalances may cause reduced productivity and adversely affect the quality of new crops planted on the reclaimed land. Salts could accumulate, and soil salinization could occur and cause reduced productivity. Because of increased exposure and porosity from strip mining, boron concentrations in surface waters and groundwater could reach levels that would damage boron-sensitive crops. Molybdenum and selenium also could be rendered more available to plants and could accumulate to such an extent in forage as to make it unsafe for animal consumption. There are insufficient data from actual field situations to make proper assessments of these potential problems. When they occur, such problems are site specific and may be mitigated through proper soil management.

The term synthetic fuels refers to liquid and gaseous fuels produced by the thermal processing of coal. Liquefaction and gasification

FIGURE 3 The scrubber system for units 1 and 2 at the Bruce Mansfield Plant is designed to remove 99.8 percent of the fly ash and 92.1 percent of the sulfur dioxide. The scrubbers require 5 percent of each unit's power output and cost approximately $13 million a year per unit to operate and maintain. (Photo courtesy Pennsylvania Power Company.)

processes for coal conversion in the United States are still in the experimental stage. There are no large-scale commercial synthetic-fuel (synfuel) plants operating within the United States, and only a few elsewhere. A review of unit operations in liquefaction and gasification processes has defined the stages where trace-element concentrations and emissions may occur. Data on the fate of trace elements in synfuel processes are beginning to emerge but are currently insufficient to provide a reliable assessment related to health.

Except for possible emissions from coal-handling facilities and process furnaces, escape of trace elements from synfuels plants should be controllable. The processing of coal in these plants occurs in closed vessels, and most of the trace elements in the feed coals will emerge in specific waste streams, which can be disposed of or treated to minimize escape of trace elements to the environment.

FIGURE 4 The gypsumlike waste from the Bruce Mansfield Plant is deposited in the valley behind the dam. The capital cost of this waste-disposal system is reported to be $90 million. (Photo courtesy Pennsylvania Power Company.)

In conclusion, the volume of chemical analytical data on U.S. coal is now considered large enough to provide the needed broad assessment of trace elements in coal and related problems. The major problem now is to identify the types of studies and possible regulatory actions needed to control those relatively few trace elements of known health concern in those particular coal beds and combustion plants that contain such trace elements in appreciable amounts. For example, arsenic and mercury are of no concern for coal mines and coal-fired power plants in the Northern Great Plains; however, close monitoring and possible regulation of arsenic in mines in the Appalachian and Interior provinces and of mercury in the Appalachian, Interior, and Gulf provinces may be necessary. Clearly, adequate information on coal is now available to use the bed-specific, site-specific approach in assessing related health hazards.

RECOMMENDATIONS

1. Determinations of the exact chemical and mineralogical form of the relatively few trace elements of health concern in specific coal beds and coal-use processes should be made (or confirmed) as soon as practical so that appropriate physical and chemical treatment measures for their removal and control can be developed.

2. Adequate information on the health effects of radioactive elements associated with coal resource development on which to make a reliable judgment should be gathered and assessed. Although available data suggest that these effects are minor, this assumption should be either substantiated or corrected by appropriate agencies.

3. A single federal agency, such as the U.S. Geological Survey with its National Coal Resources Data System, should be designated to maintain a single trace-elements-in-coal data bank, to which pertinent data would be submitted routinely by all government, university, and industry groups.

4. More data are needed on the chemical form and solubility of trace elements in the topsoil, the overburden, and the underclay associated with site-specific coals, before and after the coal is mined. These data would provide the basis for an improved prediction of mining results, particularly with respect to the new underclay-spoil interface and the porosity, density, permeability, and surface area of the materials disturbed, which will provide the basis for theoretical and empirical assessments of the effects of these changes on the mobility of trace elements. Such collections should be begun soon by both environmental and regulatory agencies.

5. More and better information on the mass balance and on the migration pathways of trace elements through major coal gasification and liquefaction processes is urgently needed; efforts to improve such data should be initiated and vigorously pursued by the appropriate agencies. Specifically, the concentrations of trace elements in hydrocarbon liquids produced from coal must be studied. The fate of these elements in subsequent uses of the liquids should also be investigated.

6. In addition to the Standard Reference Material for bituminous coal (NBS SRM 1632a) and subbituminous coal (NBS SRM 1635), standard samples of other coals, with certified critical trace-element concentrations, should be established on a priority basis by the National Bureau of Standards. Standard reference samples for lignite and anthracite are needed and should be made readily available.

7. The Panel also recommends that individual researchers in academic, industrial, and government organizations and administrators responsible for planning federal agency programs give particular attention to the following research and data needs that would improve the safe and effective exploitation of coal resources (the sequence of listing is not intended to signify relative priorities):

- Research is needed to facilitate industrial and agricultural uses of coal ash, and to identify and implement additional uses, including the development of techniques to alter the properties of the ash thus rendering it more conducive to beneficial uses.
- Data are needed on leaching of salts, boron, fluorine, molybdenum, selenium, chromium, and possibly other elements from ash piles and

lagoons. This information is needed to guard against unnecessary contamination of surface waters, groundwaters, soils, and vegetation.

• Better understanding is needed of the factors influencing the availability of molybdenum and selenium to crops grown in ash-contaminated or ash-amended soils in relation to their chemical, physical, and biological properties. Knowledge of the fate of molybdenum and selenium in ash-amended soils over extended periods of time is also needed to evaluate adequately the possible accumulations of molybdenum and selenium in forage grown on ash-amended soils to levels considered to be unsafe for livestock consumption.

• Techniques are needed to guard against excessive accumulations of salts and boron in regions adjacent to coal-ash disposal or utilization locations. Total dissolved salts and boron in leachates from coal ash frequently reach levels that are toxic to plants.

• The increased use of lime and/or limestone to remove SO_2 from flue gas will result in the production of large quantities of waste referred to as flue-gas desulfurization sludge (FGD sludge). Techniques are needed to use this sludge or to dispose of it in a manner consistent with preservation of environmental quality, particularly with relation to its readily mobilized trace elements and soluble salts derived from coal.

• Although the percentage of coal ash emitted from stacks in a modern coal-fired power plant represents less than 1 percent of the total ash that is produced, the size of particles entering the atmosphere is of submicrometer size, and they are known to have abnormally high concentrations of surface-adsorbed trace elements. Additional information is needed on the possible impact of inhalation of these submicrometer particles on human health.

• A uniform method of expressing data for the chemical composition of coal and coal ash should be established.

• Research is needed to hasten the revegetation process of renovating lands disturbed by strip mining. The use of fertilizers and irrigation to re-establish native vegetation should be investigated more thoroughly.

• Research is needed to characterize more completely the chemical composition and to develop improved techniques for the safe disposal or recycling of the large volumes of coal-cleaning liquid wastes that generally have abnormally high concentrations of certain metallic trace elements.

• Investigations are needed to determine the condition under which coal ash can be used to remove phosphorus and other undesirable trace-element constituents from surface waters and wastewaters.

• Continued efforts are required to identify regions that are subject to mobilization of trace elements by acid mine drainage (AMD) and to take appropriate corrective action, even though substantial progress toward correcting AMD has been made.

• To extend coal-ash disposal studies and to enhance evaluations of the impact of coal residues on the quality of soils, vegetation, and surface and underground waters, field-plot studies on agricultural land with coal ash containing notable amounts of possibly deleterious trace elements are needed. Such studies have been mainly limited to greenhouse conditions.

2

THE RESOURCE AND ITS UTILIZATION

COAL FORMATION

Coal, like other fossil fuels, represents the accumulation of organic materials in sedimentary strata. Unlike oil or gas, coal does not migrate, but it undergoes _in situ_ compaction and induration with time to form the various ranks of coal. Coal, which may contain recognizable source material, is composed chiefly of compressed and somewhat altered remains of terrestrial plants, such as wood, bark, roots (some still in the position of growth), leaves, spores, and seeds.

Accumulation of _in situ_ residues and imported debris (mostly inorganic sediment) in swamps leads to the formation of peat. Burial of the peat and subsequent alteration by pressure and heat convert the plant material to coal. With heat and pressure over geologic time, there is a continuous transformation from peat, made up of slightly modified plant fragments that have a substantial oxygen and hydrogen content, to hard black glistening anthracitic coal, which is predominantly carbon and contains little or no recognizable plant remains. The coal genetic series has been arbitrarily divided into a number of ranks referred to, from lowest to highest rank, as lignite, subbituminous, bituminous, and anthracite coals. The more obvious features of these ranks are shown in Table 1.

The higher the proportion of total carbon and the lower the volatile matter, the higher the rank of the coal in the series from lignite to anthracite.

For peat to undergo the transformation to coal, a series of geologic sedimentation processes must occur over time. After peat has accumulated for a period, it must be buried under mineral sediment, generally clay, silt, and sand. In time, the peat forms beds of coal that range from a few centimeters to many meters in thickness. The coal is often interbedded with shale, sandstone, and other sedimentary rocks. A single stratigraphic sequence may include several coal beds; 117 different coal beds have been identified and named in West Virginia alone. Coal-bearing strata may include alternating marine and nonmarine beds. The coal beds are in the nonmarine parts of the section and are of brackish-water or freshwater origin, though some peat swamps received occasional marine incursions. Clay and sand that washed into the swamp while the incipient coal accumulated and other inorganic components

TABLE 1 Distinctive Features and Typical Analytical Components of Coal of Various Ranks

Rank	Features[a] Physical Appearance	Characteristics
Lignite	Brown to brownish black	Poorly to moderately consolidated; weathers rapidly; plant residues apparent
Subbituminous	Black; dull or waxy luster	Weathers easily; plant residues faintly shown
Bituminous	Black; dense; brittle	Is moderately resistant to weathering; plant structures visible with microscope; burns with short blue flame
Anthracite	Black; hard; usually with glassy luster	Very hard and brittle; burns with almost no smoke

	Elemental Components[b] (percent)					
	In-Place	Dry and Ash-Free Basis				
	Moisture[c]	Carbon	Hydrogen	Nitrogen	Oxygen	Sulfur[d]
Lignite	>30	<70	>5	1-1.5	>15	0.5
Subbituminous	15-30	70-75	>5	1-1.5	10-15	0.5
Bituminous	5-15	75-91	2-5.5	1-1.5	5-10	>1.0
Anthracite	<5	>90	<2	<1	<5	0.5

[a]Source: Gilluly *et al.* (1968).

[b]The ash content is not shown because it has a wide range for all ranks of coal.

[c]Moisture refers to water held in the pore spaces within the coal.

[d]Sulfur content, like ash, depends largely on conditions that existed during accumulation of the peat and hence is independent of rank. However, in the United States the western low-rank coals tend to contain less sulfur than the higher-rank midwestern and eastern bituminous coals.

together with some remaining plant materials remain as ash when the coal is burned. Such noncombustibles lower the heating value of coal and, by increasing the waste, detract from its desirability as a fuel.

Because of the inclusion of this sediment and waterborne dissolved elements, coal beds will contain all the elements found in the eroded rocks from which the sediments are derived. A coal bed could therefore contain all the naturally occurring elements. Additional elements can be introduced into the coal after burial by infiltrating groundwater, either precipitating as minerals or through ion exchange in the clays or organic matter.

In terms of biogeochemical cycles, coal represents the fixation of carbon from atmospheric carbon dioxide by plant growth over extensive geologic time. The release of all this accumulated fixed carbon into the atmosphere as carbon dioxide may result in an unbalanced biogeochemical carbon cycle. In the long term, the consequences of this imbalance may be the most significant impact of coal utilization, and these are currently under study by many scientists. Other elements will be mobilized to a far less dramatic level but in sufficient quantities to warrant concern.

COAL USE

The higher-rank coals--anthracitic and bituminous--that occur in eastern and central North America are found in Pennsylvanian-age (upper Paleozoic) rocks (about 300 million years old). Western coals are in younger Cretaceous and lower Tertiary rocks (60 million to 135 million years old), and most are lower in rank.

Small tonnages of anthracite are mined in commercial quantities only in Pennsylvania. Younger lignitic and subbituminous coals occur principally in Montana, Wyoming, North Dakota, and the Gulf Coast states. As shown in Figure 5, coal-bearing rocks underlie about 14 percent of the land area of the contiguous United States. According to Averitt (1975), bituminous, subbituminous, lignitic, and anthracitic coals represent 43.1, 28.1, 27.7, and 1.1 percent, respectively, of the identified U.S. coal resources in 1974. The percentage of each of the different ranks of coal currently being produced in the United States and their typical heating values are shown in Table 2. This shows that more than 90 percent of the coal produced in the United States is bituminous (Swanson et al., 1976).

Coal supplied about half of the total energy requirements of the United States in 1940. In 1972, which was a record year in the production and consumption of energy in the United States, only 23 percent of the total production of energy was derived from coal, and only 19 percent of energy consumption came from coal (Averitt, 1975). Increased demand for reliable and inexpensive sources of energy and the emphasis on reducing U.S. dependence on foreign sources of fossil fuels is, however, turning the trend of coal consumption upward. The U.S. coal resources were estimated in 1974 at 3.6 trillion metric tons, an amount equivalent to 20 percent of the world's estimated total coal resources (Averitt, 1975). Geographical distribution of these reserves is shown in Figure 6 and Table 3. The 1974 estimate of the amount of coal that can be mined economically was 396 billion metric tons (U.S. Bureau of Mines, 1977).

Production of coal in 1975 was estimated at 654.6 million short tons, from which the electric utilities consumed 404.7 million tons or 61.8 percent (Table 4). Brackett (1973) estimated that by the year 1980 power utilities consumption will reach about 500 million tons annually.

Energy contained in anthracitic coals is about 1.6 times that of lignitic coals, principally because of the high moisture and oxygen contents of the low-rank coals. In coal-fired power plants, approximately 35 percent of the heat energy contained in coal is ultimately converted into electrical energy. The actual energy efficiency for a specific power plant depends on such factors as coal properties, plant size, and plant load. Larger plants are usually more efficient than smaller ones, but heat-conversion efficiency in power plants is not necessarily directly proportional to the plant size.

Because the increasing need for electrical energy is the predominant impetus for increased coal use, the present use pattern is likely to change only in degree. The use of coal to produce synthetic fuels for

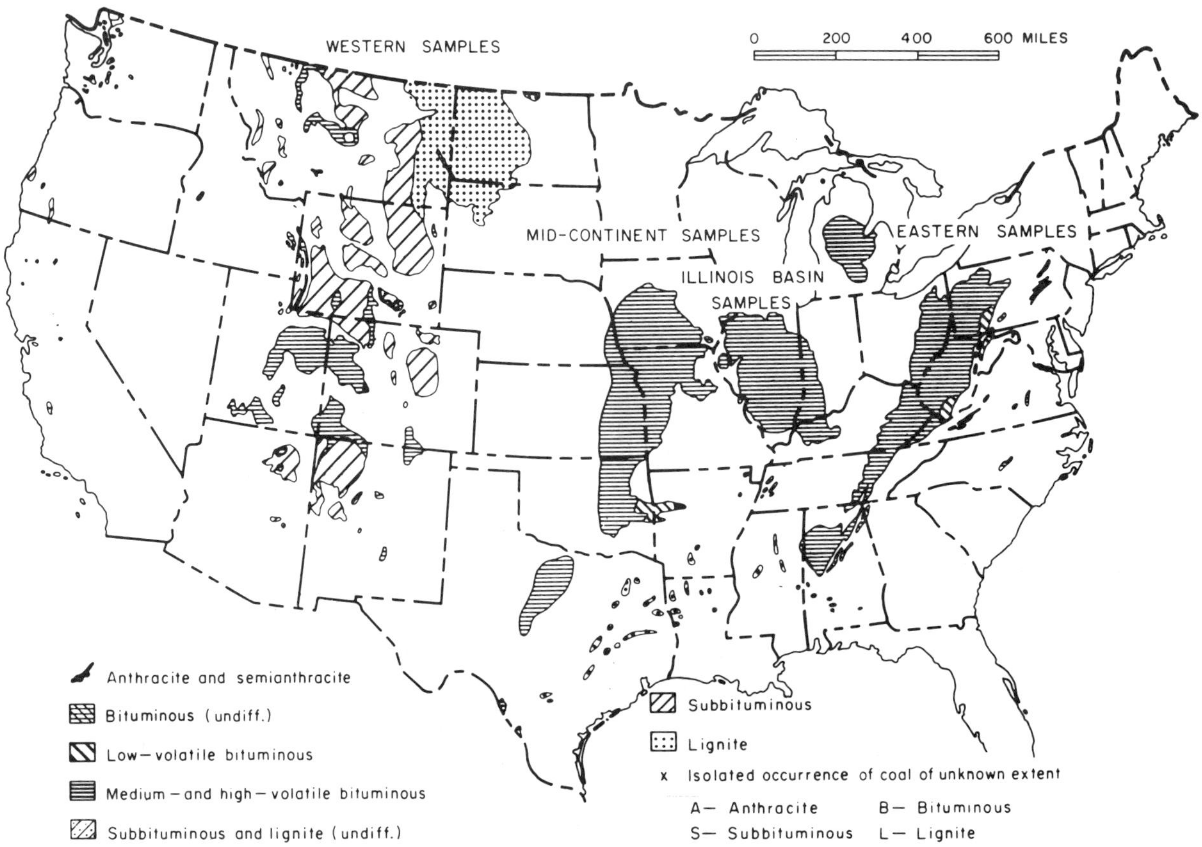

FIGURE 5 Coal fields of the conterminous United States (Gluskoter et al., 1977).

the transportation and industrial sectors will increase. Coking will probably remain at the present levels. Some increase in coal use as a chemical feedstock to replace petroleum sources is anticipated but not in the immediate future.

This report reflects the foregoing use pattern in concentrating on the mobilization of trace elements from electric power production.

TABLE 2 Proportion of Total Production and Typical Heating Values of Coal[a]

Rank	Production (percent)	Heating Value Btu/lb	Heating Value kcal/g
Anthracite	0.6	12,780	7.1
Bituminous	95.1	12,260	7.0
Subbituminous	3.1	9,410	5.2
Lignite	1.2	5,000	2.8
All coal	100	11,180	6.2

[a]Source: Swanson *et al.* (1976).

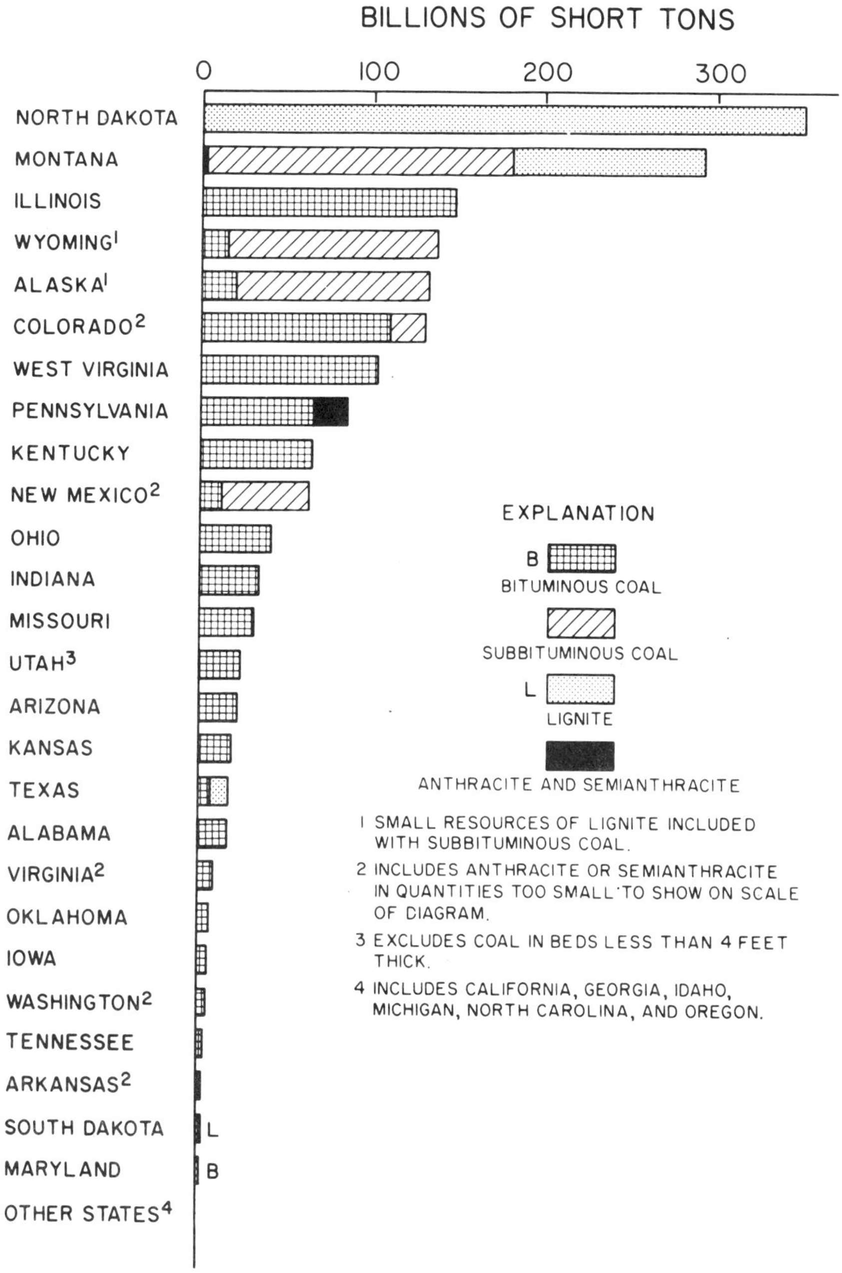

FIGURE 6 Remaining identified coal resources of the United States, January 1, 1974, by states (Averitt, 1975).

CHEMICAL AND MINERALOGICAL COMPOSITION

It should be recognized that all the material leaving the mine portal enters into the environment in one form or another. Man's objective must be to produce the material and avoid or minimize adverse environmental effects. For example, certain trace metals applied to crops as fertilizer may have a strong positive environmental effect,

TABLE 3 Total Remaining Identified Coal Resources of the United States, January 1, 1974[a] (billions of short tons)

State	Bituminous Coal	Subbituminous Coal	Lignite	Anthracite and Semianthracite	Total
Alabama	13.3	—	2.0	—	15.3
Alaska	19.4	110.7	—	—	130.1
Arizona	21.2	—	—	—	21.2
Arkansas	1.6	—	0.4	0.4	2.4
Colorado	109.1	19.7	0.02	0.08	128.9
Georgia	0.02	—	—	—	0.02
Illinois	146	—	—	—	146
Indiana	32.9	—	—	—	32.9
Iowa	6.5	—	—	—	6.5
Kansas	18.7	—	—	—	18.7
Kentucky (eastern)	28.2	—	—	—	28.2
Kentucky (western)	36.1	—	—	—	36.1
Maryland	1.2	—	—	—	1.2
Michigan	0.2	—	—	—	0.2
Missouri	31.2	—	—	—	31.2
Montana	2.3	176.8	112.5	—	291.6
New Mexico	10.8	50.6	—	0.004	61.4
North Carolina	0.11	—	—	—	0.11
North Dakota	0	—	350.6	—	350.6
Ohio	41.2	—	—	—	41.2
Oklahoma	7.1	—	—	—	7.1
Oregon	0.05	0.28	—	—	0.33
Pennsylvania	63.9	—	—	18.8	82.7
South Dakota	0	—	2.2	—	2.2
Tennessee	2.5	—	—	—	2.5
Texas	6.1	—	10.3	—	16.4
Utah	23.2	0.2	—	—	23.4
Virginia	9.2	—	—	0.3	9.5
Washington	1.9	4.2	0.1	0.005	6.21
West Virginia	100.2	—	—	—	100.2
Wyoming	12.7	123.2	—	—	135.9
Other states[b]	0.6	0.03	0.05	—	0.68
Total	747.48	485.71	478.17	19.59	1730.95

[a]Source: Averitt (1975). Some figures shown in this table were rounded off to the nearest first or second decimal place and may therefore be slightly different from the original source. Dashes indicate that tonnage is too small to make a noticeable contribution and/or is included under other ranks.

[b]Includes California, Idaho, Louisiana, and Mississippi.

whereas the same trace elements emitted to the atmosphere in an uncontrolled manner could have strong negative environmental effects.

The maximum amount of a particular element that might be mobilized into the environment from coal mining and use could be estimated from data on the total major, minor, and trace-element content of coal. By virtue of its origin, coal contains nearly every naturally occurring element. However, except for carbon, hydrogen, oxygen, nitrogen, sulfur, silicon, iron, aluminum, and the alkali and alkaline earth metals, most elements are present in coal in only minor or trace

TABLE 4 Production and Utilization of Coal in the United States, 1975[a]

	Thousands of Short Tons	Percent
Production	648,438	—
Consumption	556,301	100
Electric utilities	403,249	72.5
Coke and steel	85,987	15.5
Other manufacturers	59,783	10.8
Retail market	7,282	1.3

[a]Production and consumption figures refer to bituminous and lignite coals. Production of Pennsylvania anthracite coal amounted to 6,203 million short tons in 1975, and consumption was estimated at 5,103 million short tons. Of the amount consumed, 29 percent was used by power utilities and 41.6 percent was consumed in industrial and residential heating. Data were combined from U.S. Bureau of Mines (1977).

amounts. Swaine (1977) indicated that some trace elements in coal may be associated with phenolic, carboxylic, amide, and sulfhydroxyl functional groups in the organic fraction of coal. In the inorganic fraction, the trace elements are often associated with clays, silicates, carbonates, sulfides, sulfates, and other minerals. Table 5 shows the association of some trace elements with mineral matter. Layer silicate minerals (clays), quartz, and pyrite are widely distributed in coals. Kaolinite is the most common layered silicate, although montmorillonite and illite are also found. Dolomite, siderite, calcite, and aragonite are the most common carbonate minerals. Pyritic and marcasitic crystalline forms of FeS_2 are by far the dominant sulfide minerals occurring in coals of the United States. Barite, zircon, and fluorapatite occur sporadically.

The concentration of inorganic elements in a given coal deposit depends in part on the conditions that prevailed in the peat swamp in which the coal-forming material was deposited and on the subsequent genetic history of the coal seam. These conditions, especially the hydrological conditions that affected the transport of inorganics into the swamp, varied geographically and temporally within a given peat swamp. Therefore, the concentration of inorganic elements, as both mineral matter and organically combined species, varies from one location to another in a coal bed and from one bed to another. Typical compositional values are shown in Table 6, but it should be clearly understood that any given coal seam may contain concentrations of specific elements that deviate significantly from these values.

Silicon, aluminum, calcium, magnesium, sodium, potassium, iron, manganese, and titanium are the major inorganic elements of coals but generally account for <10 percent of its total elemental composition. Lignitic coals generally contain higher amounts of silicon, calcium, magnesium, and sodium than do other ranks of coal. Of the trace and

TABLE 5 Possible Mineral Associations of Trace Elements with Mineral Matter in Coal[a]

Element	Mineral Association
Arsenic	Arsenopyrite (FeAsS)
Barium	Barite ($BaSO_4$)
Boron	Illite and tourmaline (complex aluminum silicates)
Cadmium	Sphalerite [(Zn,Cd)S]
Cobalt	Linnaeite (Co_3S_4)
Copper	Chalcopyrite ($CuFeS_2$)
Fluorine	Fluorapatite [$Ca_5(PO_4)_3(F, OH)$]
Lead	Galena (PbS)
Manganese	Siderite [$(Fe,Mn)CO_3$] and calcite [$(Ca,Mn)CO_3$]
Mercury	Pyrite (FeS_2)
Molybdenum	Molybdenite (MoS_2)
Nickel	Millerite (NiS)
Phosphorus	Fluorapatite [$Ca_5(PO_4)_3(F, OH)$]
Selenium	Pyrite (FeS_2)
Strontium	Goyazite group (hydrous strontium aluminum phosphates)
Zinc	Sphalerite (ZnS)
Zirconium	Zircon ($ZrSiO_4$)

[a]Source: Swaine (1977).

minor elements shown in Table 6, only mercury, lead, molybdenum, arsenic, cadmium, boron, antimony, and selenium are generally richer in coal than in the average composition of the Earth's crust (Table 7). For comparison, Table 6 also shows that the estimated average inorganic-element concentrations of coal on a worldwide basis are, in general, in the range of those for U.S. coals.

The concentrations of most of the elements that occur in U.S. coal were compiled by the U.S. Geological Survey from analyses of 799 samples (Swanson et al., 1976). Their compilation is extracted in Table 8. The analyses presented in Tables 7 and 8 are on a whole-coal basis. As of 1979, these analyses and similar analyses on 3700 coal samples from the United States are stored in and available from the data bank of the U.S. Geological Survey in the National Coal Resources Data System (Carter, 1976).

Regional Variations and Concentration Patterns

Examination of Table 8 shows the variation that occurs in the elemental compositions of coals obtained from a number of coal-production locations in the United States. Elements considered by the Panel to be

TABLE 6 Chemical Composition of U.S. Coals[a,b]

Element	Anthracite	Bituminous	Sub-bituminous	Lignite	Average	Estimated Worldwide Average[c]
Major elements (percent)						
Sulfur (S), total	0.8	2.7	0.7	1.7	2.0[d]	2.0
Sulfur (S), sulfate	0.02	0.16	0.04	0.24	0.12	—
Sulfur (S), pyritic	0.35	1.70	0.36	0.68	1.19	—
Sulfur (S), organic	0.48	0.88	0.32	0.75	0.70	—
Phosphorus (P)	—	—	0.10	0.007	—	0.05
Silicon (Si)	2.7	2.6	2.0	4.9	2.6	2.8
Aluminum (Al)	2.0	1.4	1.0	1.6	1.4	1.0
Calcium (Ca)	0.07	0.33	0.78	1.2	0.54	1.0
Magnesium (Mg)	0.00	0.08	0.18	0.31	0.12	0.02
Sodium (Na)	0.05	0.04	0.10	0.21	0.06	0.02
Potassium (K)	0.24	0.21	0.06	0.20	0.18	0.01
Iron (Fe)	0.44	2.2	0.52	2.0	1.6	1.0
Manganese (Mn)	0.002	0.01	0.006	0.015	0.01	0.005
Titanium (Ti)	0.15	0.08	0.05	0.12	0.08	0.05
Minor and trace elements (ppm)						
Antimony (Sb)	0.9	1.4	0.7	0.7	1.1	3.0
Arsenic (As)	6	25	3	6	15	5.0
Barium (Ba)	100	100	300	300	150	500
Beryllium (Be)	1.5	2.0	0.7	2.0	2.0	3
Bismuth (Bi)	—	—	<0.6	<0.8	<0.7	5.5
Boron (B)	10	50	70	100	50	75
Bromine (Br)	—	—	2.3	3.2	2.6	—
Cadmium (Cd)	0.3	1.6	0.2	1.0	1.3	—
Cerium (Ce)	—	—	5.5	12.3	7.7	11.5
Cesium (Cs)	—	—	0.40	0.25	0.4	—
Chlorine (Cl)	—	—	255	110	207	1000
Chromium (Cr)	20	15	7	20	15	10
Cobalt (Co)	7	7	2	5	7	5
Copper (Cu)	27	22	10	20	19	15
Dysprosium (Dy)	—	—	2.7	1.4	2.2	—
Erbium (Er)	—	—	0.46	0.16	0.34	0.6
Europium (Eu)	—	—	0.61	0.13	0.45	0.7
Fluorine (F)	61	77	63	94	74	—
Gadolinium (Gd)	—	—	0.13	0.21	0.17	1.6
Gallium (Ga)	7	7	3	7	7	7
Germanium (Ge)	—	—	0.47	1.20	0.71	5
Hafnium (Hf)	—	—	0.75	0.30	0.60	-
Holmium (Ho)	—	—	0.13	0.06	0.11	0.3
Iodine (I)	—	—	1.0	1.3	1.10	—
Lanthanum (La)	—	—	7.5	3.2	6.1	10
Lead (Pb)	10	22	5	14	16	25
Lithium (Li)	33	23	7	19	20	65
Lutetium (Lu)	—	—	0.09	0.05	0.08	0.07
Mercury (Hg)	0.15	0.20	0.12	0.16	0.18	0.012
Molybdenum (Mo)	2	3	1.5	2	3	5
Neodymium (Nd)	—	—	50	11	37	4.7
Nickel (Ni)	20	20	5	15	15	15
Niobium (Nb)	—	—	5.4	2.7	4.5	—

TABLE 6 (continued)

Element	Anthracite	Bituminous	Sub-bituminous	Lignite	Average	Estimated Worldwide Average[c]
Praseodymium (Pr)	—	—	6.1	2.7	2.7	2.2
Rubidium (Rb)	—	—	5.3	0.98	2.90	100
Samarium (Sm)	—	—	0.50	0.27	0.42	1.6
Scandium (Sc)	5	3	2	5	3	5
Selenium (Se)	3.5	4.6	1.3	5.3	4.1	3
Silver (Ag)	—	—	0.17	0.26	0.20	0.50
Strontium (Sr)	100	100	100	300	100	500
Tellurium (Te)	—	—	0.1	0.1	0.1	—
Terbium (Tb)	—	—	0.1	0.1	0.1	0.3
Thallium (Tl)	—	—	0.1	0.12	0.1	—
Thorium (Th)	—	—	2.3	1.2	1.9	—
Thulium (Tm)	—	—	0.07	0.07	0.07	—
Tin (Sn)	—	—	1.7	1.3	1.6	—
Tungsten (W)	—	—	2.2	3.3	2.5	—
Uranium (U)	1.5	1.9	1.3	2.5	1.6	1.0
Vanadium (V)	20	20	15	30	20	25
Ytterbium (Yb)	1	1	0.5	1.5	1	0.5
Yttrium (Y)	10	10	5	15	10	10
Zinc (Zn)	16	53	19	30	39	50
Zirconium (Zr)	50	30	20	50	30	—

[a] Generalized from Swanson *et al.* (1976).

[b] Dashes indicate no data included in source.

[c] Generalized from Bertine and Goldberg (1971) and U.S. Environmental Protection Agency (1975).

[d] Except for sulfur, the average composition of U.S. coals was calculated from the analysis of 53, 509, 183, and 54 samples of anthracite, bituminous, subbituminous, and lignite, respectively; the all-coal average is calculated from analysis of 799 samples. The average sulfur was calculated from the analysis of 38, 277, 105, and 26 samples of these coals, respectively. The all-coal average sulfur is from the analysis of 488 samples. See footnotes to Table 8 for methods used for determining mean concentrations.

TABLE 7 Enrichment Factors of Trace Elements in Coal with Respect to the Average Composition of the Earth's Crust[a,b]

Limited Enrichment		Moderate Enrichment		Marked Enrichment	
Element	Enrichment	Element	Enrichment	Element	Enrichment
Hg	2.25	As	8.3	Se	82
Mo	2.00	Cd	6.5		
Pb	1.28	B	5.0		
		Sb	5.5		

[a] Refer to coal composition in Table 6.

[b] The average chemical composition of the Earth's crust was taken from data reported by Taylor (1964).

TABLE 8 Average U.S. Coal Composition by Location[a,b]

Element[c]	Appalachian Region	Interior Province	Gulf Province	Northern Great Plains Province	Rocky Mountain Province	Alaska
Major elements (percent)						
Sulfur, total	2.3	3.9	1.9	1.2	0.6	0.2
Sulfur, sulfate	0.09	0.27	0.33	0.03	0.05	0.01
Sulfur, pyritic	1.56	2.37	0.59	0.76	0.19	0.07
Sulfur, organic	0.74	1.25	0.96	0.37	0.32	0.12
Nitrogen	1.3	1.2	0.4	0.9	1.2	0.7
Silicon	2.7	2	6.6	1.4	2.5	2.9
Aluminum	1.6	0.97	2.1	0.69	1.2	1.5
Calcium	0.12	1.2	1.2	0.97	0.59	1
Magnesium	0.068	0.089	0.291	0.255	0.104	2.5
Sodium	0.032	0.035	0.732	0.182	0.102	0.018
Potassium	0.23	0.16	0.3	0.04	0.076	0.12
Iron	1.9	3.3	2.2	0.75	0.45	0.38
Manganese	0.062	0.014	0.024	0.0051	0.0036	0.0061
Titanium	0.09	0.052	0.16	0.042	0.061	0.077
Minor and trace elements (ppm)						
Antimony	1.2	1.7	0.9	0.6	0.4	2.7
Arsenic	27	21	6	3	2	3
Barium	100	70	200	500	200	700
Beryllium	2	3	2	0.5	0.7	0.7
Boron	30	100	100	70	70	70
Cadmium	0.7	7.1	1.3	0.2	<0.1	<0.2
Chromium	20	15	20	5	5	15
Cobalt	7	7	7	2	2	5
Copper	24	20.2	28	8.3	9.1	16.8
Fluorine	80	71	124	45	70	90
Gallium	7	5	10	3	3	5
Lead	15.3	55	20	5.3	5.5	5.9
Lithium	27.6	11	28	6.0	9.2	10.1
Mercury	0.24	0.14	0.18	0.09	0.06	4.4
Molybdenum	3	5	3	2	1.5	1.5
Nickel	15	30	20	3	30	10
Niobium	5	1.5	7	5	1	3
Scandium	5	3	7	2	2	5
Selenium	4.7	4.6	7	1.0	1.6	2
Strontium	100	50	200	150	100	100
Thallium	4.9	5.2	8.3	2.7	3.6	4.4
Uranium	1.4	3.3	3.2	0.9	1.6	1.2
Vanadium	20	20	50	10	15	30
Ytterbium	1	0.7	2	0.3	0.5	1
Yttrium	10	10	20	5	5	10
Zinc	20	373	40	25.6	9.9	24
Zirconium	50	15	70	15	20	20

[a]Source: Swanson *et al.* (1976).

[b]Sulfur and nitrogen analyses are expressed on an as-received basis. Arsenic, fluorine, mercury, antimony, selenium, thorium, and uranium were determined directly on a whole-coal basis. The remaining elements were determined in the ash and are expressed on a whole-coal basis.

[c]Average sulfur and nitrogen calculated from analysis of 158, 90, 19, 40, 86, and 9 samples from Appalachian region, Interior province, Gulf province, Northern Great Plains province, Rocky Mountain province, and Alaska, respectively. Averages of the remainder of the elements were calculated from the analysis of 331, 143, 34, 93, 124, and 18 samples from the Appalachian region, Interior province, Gulf province, Northern Great Plains province, Rocky Mountain province, and Alaska, respectively.

of greatest concern (arsenic, boron, cadmium, fluorine, lead, mercury, molybdenum, and selenium) range in average concentration in coals from the regional highs and lows shown in Table 9.

Table 8 also shows large variations in the concentrations of those elements considered to be of moderate concern (chromium, vanadium, copper, zinc, and nickel) among coals from those regions. These data demonstrate the necessity of using the site-specific approach in assessing health-related hazards associated with trace elements in coal resource development in the United States.

Gluskoter et al. (1976, 1977), at the Illinois State Geological Survey, have conducted a study that further defines the patterns of variability of trace-element concentrations within coal. They found that those elements with the greatest variability are associated with sulfide minerals in coal. Elements that occur in organic combinations or that are contained within the silicate minerals have the narrowest ranges of concentrations. Elemental concentrations tend to be higher in eastern coals, lower in western coals, and intermediate in the Interior coal province. The concentrations of many trace elements are positively correlated with each other in their occurrence. The highest correlation was between zinc and cadmium concentrations. The concentrations of chalcophile elements (antimony, arsenic, cobalt, lead, and nickel) are also mutually correlated, as are calcium with manganese and sodium with chlorine. The Gluskoter study consistently found the selenium content of coal to be enriched compared with the average for the earth's crust.

In earlier work at the Illinois State Geological Survey by Ruch et al. (1974), it was found that certain of the elements (iron, beryllium, boron, chromium, copper, nickel, selenium, and vanadium) display a relatively normal distribution of concentration among samples; others (cadmium, zinc, arsenic, antimony, lead, tin, and mercury) exhibit a skewed and wide range of concentrations. The geochemical relations and associations of trace elements in coal have also been reviewed by Zubovic (1976).

The variability in coal composition is large, and the mobilization from its use will largely depend on the coal, mining methods, type of transport (rail, barge, pipeline, or truck), and energy or industrial conversion process. Any harmful consequence of the mobilization will be

TABLE 9 Highest and Lowest Concentration (and Region) of the Elements of Greatest Concern to the Panel[a]

Element	Low		High	
	Concentration	Province	Concentration	Province
Arsenic	2	Rocky Mountain	27	Appalachian
Boron	30	Appalachian	100	Gulf and Interior
Cadmium	<0.2	Alaska	7.1	Interior
Lead	5.3	Northern Great Plains	55	Interior
Mercury	0.06	Rocky Mountain	4.4	Alaska
Molybdenum	1.5	Rocky Mountain and Alaska	5	Interior
Selenium	1	Northern Great Plains	4.7	Appalachian

[a]Source: Table 8.

dependent on the environment into which the elements are released. In this analysis, all possibilities cannot be considered, and an attempt has been made to look at the problem "on the average" as well as a few specific problems.

Uranium in Coal

Uranium is present in most coal in about the same very small amounts as it is in most rocks and soils the world over--a few parts per million. Such concentrations that apply to approximately 97 percent of the coal mined and burned in the United States (Swanson et al., 1976) are based on many thousands of analyses for uranium in coal throughout the country. Therefore, rather than be improperly cited as a major or specific cause for environmental concern, uranium in coal should be clearly relegated to a category of minimal concern (Farmer et al., 1977; Peyton, 1976; Stanford Research Institute, 1977).

What little uranium there is in most coal is almost entirely concentrated in the bottom ash during combustion. It is not volatilized; rather, as an exceptionally heavy element, it reacts chemically and physically as a refractory element. The uranium, which is disseminated in the coal, becomes firmly bound in the glassy ash and is released only in very long-term reaction with natural waters. It is released from the stacked or buried ash in the same extremely minor amounts that are slowly released and dissolved from most soils and unconsolidated sediments. The relatively minor amount of uranium that is in the fly ash (generally <0.0002 percent, or <2 ppm), which is produced as particulate exhaust from a combustion plant, is so small that it can be considered as actually diluting the uranium content of the soil or other surface material on which it comes to rest.

Coal in only two areas of the United States is known to contain abnormally high amounts of uranium. That in the north-central Midcontinent is high-sulfur bituminous coal of Pennsylvanian age, used principally in a few relatively small electricity-generation plants in Missouri and southern Iowa. That in the second area is spottily distributed uraniferous lignite near the northwestern corner of South Dakota. This lignite is a very high-ash low-Btu lignite of Paleocene age that is not now of economic value. The lignite was strip mined locally for uranium in 1957-1962; the uranium was concentrated in the ash by spreading the lignite on the ground and waste-burning the combustible fraction. The uranium content of the north-central Midcontinent bituminous coal is generally 10 to 15 ppm; the sparsely distributed "uranium lignite" of South Dakota rarely contains more than 100 ppm.

To release and recover the uranium from the ash of the combusted coal in both areas would require the pulverization of the ash, followed by treatment with strong acid and alkali chemicals, a process that is far too expensive (equipment- and labor-intensive) to be considered economical even with current market values of $30 to $40 per pound of U_3O_8.

As for each of the many trace elements that are potentially dangerous to health, no matter how they may be artificially released to the environment, the uranium content of coal and other disturbed rock units at each planned coal mine should be precisely determined. For those few mines and coal-using plants where uranium content of the coal is as much as three times above that of the Earth's average of about 2 ppm, special investigations should be required to define the rates, amounts, routes of travel, and long-term residence sites of such uranium.

Relatively little is known about the many so-called radionuclides in coal, although some have various degrees of volatility, and some are associated with or are decay products of uranium. It is known, however, that during coal usage radionuclides are produced in almost infinitesimal amounts, barely detectible with modern mass-spectrographic instruments. The several existing reports on radionuclides in coal are cited in this report, and the minimal information and the need for additional work are clearly specified in the report.

ANALYTICAL PROCEDURES AND AVAILABLE STANDARDS FOR COAL AND COAL-RESIDUE ANALYSIS

In recent years, general interest in chemical analysis of coal and coal-related materials has significantly increased to the state where adequate analytic procedures and suitable standards are currently available for many trace elements. This progress has been possible only because many governmental and private laboratories made a concerted, cooperative effort based on anticipated needs of the coal industry.

The number of complete modern analyses for trace elements in coal has increased greatly in recent years. As of 1979, the largest accumulation of these data, on 3700 samples of U.S. coal, is publicly available from the data bank of the U.S. Geological Survey in its National Coal Resources Data System (Carter, 1976).

Although much progress on chemical analysis of coal has occurred in recent years, much remains to be accomplished on such coal-related materials as conversion products, intermediates, and wastes.

Chemical Procedures

Over the years a host of techniques has evolved for determining trace elements in coal, coke, and coal ash. More recently, the trend has been to develop instrumental techniques, in preference to chemical, colorimetric, or "wet" techniques, because these procedures tend to be faster, less tedious, and generally yield acceptable results when sufficiently checked.

Several comprehensive reviews and bibliographies (Averitt *et al*., 1972, 1976; Freedman and Sharkey, 1972; Gluskoter *et al*., 1980; Ignasiak *et al*., 1975; Konieczynski, 1969) detail the available literature. Perhaps the best guide to the determination of specific elements in coal is the bi-yearly application review on "Solid and Gaseous Fuel" (Hattman *et al*., 1977). Recent symposia of the American Chemical Society Fuel Division have dealt with the analysis of coal for trace elements (Babu, 1975; American Chemical Society, 1977a). Several organizations have published individual summaries of procedures used in their laboratories

(Ford *et al.*, 1976; Gluskoter *et al.*, 1977; Karr, 1978; Pollock, 1975; Swanson and Huffman, 1976).

Coal ash and coal-related materials such as conversion wastes are also readily analyzed for trace elements by several techniques. Recent work (Sather *et al.*, 1975) concerned with the analysis of conversion ash materials indicates that excellent agreement between two different laboratories using two or more different methods can be obtained. Mass balance studies concerned with liquefaction (Filby *et al.*, 1976; Schultz *et al.*, 1977) and gasification (Attari, 1973; Forney *et al.*, 1976) processes indicate that a high level of analytical capability is available for application to these difficult matrixes, which range from dominantly organic to dominantly silicate materials.

The present trend is toward development of multielemental instrumental procedures, to quantitatively cover as many elements as are applicable. Because any particular analytical discipline is better suited for certain elements than for others for various inherent reasons, a combination of methods is usually necessary to determine all elements of interest. For example, instrumental neutron activation analysis (INAA) is based on the detection of induced radioactivity. Detection is dependent on several nuclear factors and elemental concentrations relative to other possible interfering elements. For such a complex matrix as coal, coal ash, or residues, about 40 elements are currently detectable with varying degrees of acceptable accuracy and precision. Other analytical disciplines (such as optical emission and atomic absorption spectroscopy) have corresponding principles, limitations, and areas of application. The combined use of analytical disciplines thus allows overlap, enabling an approach to better accuracy. Many laboratories successfully use this approach.

Table 10 lists some of the significant literature available for the determination of specific trace elements in coal and coal ash. One section lists those references concerned primarily with multielemental approaches, which include many or all of the individual elements listed. No attempt is made here to classify any method as superior to another because each, when properly developed and used, has definite inherent advantages. Currently, of those listed in Table 10, the most-used methods for analyzing trace elements in coal are activation analysis, optical emission, atomic absorption, x-ray fluorescence, and mass spectrometry or combinations of these methods. In general, each of these methods is quantitatively applicable to many elements, not just for those elements listed. Mass spectrometry, unless precise and accurate isotopic dilution techniques are employed, is considered to be only semiquantitative. Most colorimetric procedures, although quantitative, are usually specific only for a particular element, hence limiting. Because of the many analytical problems incurred with complex matrixes, polarography has not been extensively developed and applied to coal analysis.

Those methods generally offering lower detection limits are atomic absorption and neutron activation, usually of the order of 1 ppm or better for applicable elements. In the 1- to 10-ppm range and above, x-ray fluorescence and optical emission methods are generally useful for many elements. All of these methods are capable of acceptable precision and accuracy (e.g., $\pm$10-15 percent of the true value) for the determination of trace elements at concentrations significantly above

TABLE 10 Literature Available for the Determination of Trace Elements in Coal and Coal Ash[a]

	Multi-elemental	Sb	As	Be	B	Cd	Cr	Cu	Fe	Pb	Mn	Hg	Ni	Se	V	Zn	Th	U
Activation analysis	1-12	20	37	—	—	—	—	—	67	—	—	75, 76	—	87	—	—	95, 96, 107	95, 96, 98, 99, 107
Optical emission	13-19	—	38, 39	50	55-58	62	—	—	—	—	73	—	—	88	90	62	—	—
Atomic absorption	20-22	—	40	51	—	63	64	—	—	71	—	77-83	—	—	—	63	—	—
X-ray fluorescence	23-27	—	—	—	—	—	—	—	—	—	—	—	—	89	—	—	97	97
Mass spectrometry	28-30	—	—	52	—	—	—	—	—	—	—	—	—	—	—	—	—	—
Chemically based (includes colorimetric and polarographic	31	—	41-49	53, 54	59-61	—	65	66	68-70	72	74	84-86	66	—	65, 91	92-94	—	100-106
Mixed approaches	20, 32-36	—	—	—	—	—	—	—	—	—	—	—	—	—	—	—	—	—

[a]Numbers are keyed to accompanying references.

Table 10 (Continued)
Literature Available for the Determination of
Trace Elements in Coal and Coal Ash

1. K.H. Abel and L.A. Rancitelli. 1975. Major, minor, and trace element composition of coal and coal ash. In Trace Elements in Fuel, S.P. Babu, ed. Advances in Chemistry Series No. 141, American Chemical Society, Washington, D.C., pp. 118-138.
2. C. Block and R. Dams. 1975. Inorganic composition of Belgian coals and coal ashes. Environ. Sci. Technol. 9(2):146-150.
3. C. Block and R. Dams. 1973. Determination of trace elements in coal by instrumental neutron activation analysis. Anal. Chim. Acta 68(1):11-24.
4. J.K. Frost, P.M. Santoliquido, L.R. Camp, and R.R. Ruch. 1975. Trace elements in coal by neutron activation analysis with radiochemical separations. In Trace Elements in Fuel, S.P. Babu, ed. Advances in Chemistry Series No. 141, American Chemical Society, Washington, D.C., pp. 84-97.
5. G.E. Gordon. 1971. Instrumental activation analysis of atmospheric pollutants and pollution source materials. In Proceedings Conf-710668, Identification and Measurement of Environmental Pollutants, B. Westley, ed. National Research Council of Canada, Ottawa, pp. 138-143.
6. W.S. Lyon and J.F. Emery. 1975. Neutron activation analysis applied to the study of elements entering and leaving a coal-fired steam plant. Int. J. Environ. Anal. Chem. 4(2):125-133.
7. H.T. Millard and V.E. Swanson. 1975. Neutron activation analysis of coals using instrumental techniques. Am. Nucl. Soc. Trans. 21:108-109.
8. E. Orvini, T.E. Gills, and P.D. LaFleur. 1974. Method for determination of selenium, arsenic, zinc, cadmium, and mercury in environmental matrices by neutron activation analysis. Anal. Chem. 46:1294-1297.
9. R.A. Nadkarni. 1975. Multi-element analysis of coal and coal fly-ash standards by instrumental neutron activation analysis. Radiochem. Radioanal. Lett. 21(3-4):161-176.
10. R.R. Ruch, R.A. Cahill, J.K. Frost, L.R. Camp, and H.J. Gluskoter. 1975. Trace elements in coals of the United States determined by activation analysis and other techniques. Am. Nucl. Soc. Trans. 21:107-108.
11. D.W. Sheibley. 1975. Trace elements by instrumental neutron activation analysis for pollution monitoring. In Trace Elements in Fuel, S.P. Babu, ed. Advances in Chemistry Series No. 141, American Chemical Society, Washington, D.C., pp. 98-117.
12. W.H. Zoller. 1973. Photon activation analysis. Proceedings of 2nd Joint Conf. Sensing Environ. Pollutants, Pittsburgh, Pa., pp. 185-189.
13. R.F. Abernethy, M.J. Peterson, and F.H. Gibson. 1969. Spectrochemical analyses of coal ash for trace elements. U.S. Bur. Mines Rep. Invest. 7281. 20 pp.
14. G.B. Dreher and J.A. Schleicher. 1975. Trace elements in coal by optical emission spectroscopy. In Trace Elements in Fuel, S.P.

Table 10 (Continued)
Literature Available for the Determination of
Trace Elements in Coal and Coal Ash

Babu, ed. Advances in Chemistry Series No. 141, American Chemical Society, Washington, D.C., pp. 35-47.

15. R.L. O'Neil and N.W. Suhr. 1960. Determination of trace elements in lignite ashes. Appl. Spectrosc. 14:45-50.

16. M.J. Peterson and J.B. Zink. 1964. A semiquantitative spectrochemical method for analysis of coal ash. U.S. Bur. Mines Rep. Invest. 6496. 15 pp.

17. W. Radmacher and H. Hessling. 1959. Trace elements in coal and their spectroanalytical determination. Z. Anal. Chem. 167:172-182.

18. A.C. Smith. 1958. The determination of trace elements in pulverized-fuel ash. J. Appl. Chem. (London) 8:636-645.

19. C. Luedke and G. Holdt. 1971. Spectral analytical determination of trace contents in lignite-tar pitch coke. Izv. Fiz. Inst. ANEB (At. Nauchnoeksp. Bazc), Bulg. Akad. Nauk 21:109-114.

20. R.R. Ruch, H.J. Gluskoter, and N.F. Shimp. 1974. Occurrence and distribution of potentially volatile trace elements in coal: A final report. Ill. State Geol. Surv. Environ. Geol. Notes 72. 96 pp.

21. C.T. Ford, R.R. Care, and R.E. Bosshart. 1976. Preliminary evaluation of the effect of coal cleaning on trace element removal. Trace Element Program, Report 3. Bituminous Coal Research, Inc., Monroeville, Pa. 115 pp.

22. R.B. Muter and C.F. Cockrell. 1969. Determination of sodium, potassium, calcium, and magnesium in siliceous coal ash and related materials by atomic-absorption spectrophotometry. Appl. Spectrosc. 23:493-496.

23. M. Berman and S. Ergun. 1968. Analysis of mineral matter in coals by x-ray fluorescence. U.S. Bur. Mines Rep. Invest. 7124.

24. J.K. Kuhn, W.F. Harfst, and N.F. Shimp. 1975. X-ray fluorescence analysis of whole coal. In Trace Elements in Fuel, S.P. Babu, ed. Advances in Chemistry Series No. 141, American Chemical Society, Washington, D.C., pp. 66-73.

25. C.J. Sparks, Jr., O.B. Cavin, L.A. Harris, and J.C. Ogle. 1973. Simple quantitative x-ray fluorescent analysis for trace elements. In Proceedings of the Seventh Annual Conference on Trace Substances in Environmental Health, June 12-14, 1973, D.D. Hemphill, ed. University of Missouri, Columbia, pp. 361-368.

26. P.C. Simms, F.A. Ricky, and K.A. Mueller. 1976. Trace element analysis of coal using proton induced x-ray emission. Progress on Coal-Related Research at Purdue, May 9. Purdue University, pp. 9-12.

27. R.L. Walter, R.D. Willis, W.F. Gutknecht, and J.M. Joyce. 1974. Analysis of biological, clinical, and environmental samples using proton-induced x-ray emission. Anal. Chem. 46(7):843-855.

28. J.A. Carter, R.L. Walker, and J.R. Sites. 1975. Trace impurities in fuels by isotope dilution mass spectrometry. In Trace Elements in Fuel, S.P. Babu, ed. Advances in Chemistry Series No. 141, American Chemical Society, Washington, D.C., pp. 74-83.

Table 10 (Continued)
Literature Available for the Determination of
Trace Elements in Coal and Coal Ash

29. A.G. Sharkey, Jr., T. Kessler, and R.A. Friedel. 1975. Trace elements in coal dust by spark-source mass spectrometry. In Trace Elements in Fuel, S.P. Babu, ed. Advances in Chemistry Series No. 141, American Chemical Society, Washington, D.C., pp. 48-56.
30. R. Brown, M.L. Jacobs, and H.E. Taylor. 1972. Survey of the most recent applications of spark source mass spectrometry. Am. Lab. 4(11):29-40.
31. M. Weclewska. 1960. Polarographic determination of trace metals in coal ash. Prace Glownego Inst. Gornictwa, Komun. (Katowice) Ser. B. 242, pp. 1-12.
32. N.E. Bolton, J.A. Carter, J.F. Emery, C. Feldman, W. Fulkerson, L.D. Hulett, and W.S. Lyon. 1975. Trace element mass balance around a coal-fired steam plant. In Trace Elements in Fuel, S.P. Babu, ed. Advances in Chemistry Series No. 141, American Chemical Society, Washington, D.C., pp. 175-187.
33. J.W. Kaakinen, R.M. Jorden, M.H. Lawasani, and R.E. West. 1975. Trace element behavior in a coal-fired power plant. Environ. Sci. Technol. 9(9):862-869.
34. V.S. Tikhonova, M.Yu. Fedushchak, and S.B. Kazakov. 1973. Rare and trace elements in coals of the Lvov-Volyn Basin. Geol. Geokhim. Goryuch. Iskop. 34:104-107.
35. E.N. Pollock. 1975. Trace impurities in coal by wet chemical methods. In Trace Elements in Fuel, S.P. Babu, ed. Advances in Chemistry Series No. 141, American Chemical Society, Washington, D.C., pp. 23-34.
36. H.J. Gluskoter, R.R. Ruch, W.G. Miller, R.A. Cahill, G.B. Dreher, and J.K. Kuhn. 1977. Trace elements in coal: occurrence and distribution. Ill. State Geol. Surv. Circ. 499. Urbana, Ill. 154 pp.
37. P.M. Santoliquido. 1973. Use of inorganic ion exchangers in the neutron activation determination of arsenic in coal ash. Radiochem. Radioanal. Lett. 15(6):373-377.
38. R.H. Hall and H.L. Lovell. 1958. Spectroscopic determination of arsenic in anthracite coal ashes. Anal. Chem. 30:1665-1669.
39. N.A. Kekin and V.E. Marincheva. 1970. Spectrographic determination of arsenic in coal and coke. Zavod. Lab. 36:1061-1063.
40. G.I. Spielholtz, G.C. Toralballa, and R.J. Steinberg. 1971. Determination of arsenic in coal and in insecticides by atomic absorption spectroscopy. Mikrochim. Acta 6:918-923.
41. R.F. Abernethy and F.H. Gibson. 1968. Colorimetric method for arsenic in coal. U.S. Bur. Mines. Rep. Invest. 7184. 10 pp.
42. R.G. Ault. 1961. Determination of arsenic in malting coal. J. Inst. Brewing 67(1):14-28.
43. British Standards Institute. 1960. Methods for the analysis and testing of coal and coke. BS 1016, Part 10, Arsenic in Coal and Coke. British Standards Institute, London. 22 pp.
44. A. Crawford, J.G. Palmer, and J.H. Wood. 1958. The determination of arsenic in microgram quantities in coal and coke. Mikrochim. Acta 1958:277-294.

Table 10 (Continued)
Literature Available for the Determination of
Trace Elements in Coal and Coal Ash

45. S.R. Crook and S. Wald. 1960. The determination of arsenic in coal and coke. Fuel 39:313-322.
46. L.J. Edgcombe and H.K. Gold. 1955. Determination of arsenic in coal. Analyst 80:155-157.
47. E. Jackwerth and H.G. Kloppenburg. 1962. X-ray fluorescence spectroscopic determination of trace arsenic in conjunction with a modified gutzeit method. Z. Anal. Chem. 186:428-436.
48. F.H. Kunstmann and L.B. Bodenstein. 1961. Arsenic content of South African coals. J. S. Afr. Inst. Min. Metall. 62:234-244.
49. A.D. Wilson and D.T. Lewis. 1963. Application of the uranyl salt method to the determination of arsenic by the oxygen flask technique. Analyst 88:510-515.
50. O.A. Kul'skaya and O.F. Vdovenko. 1961. Spectrochemical determination of germanium, beryllium, and scandium in coal ash. Khim. Fis. Khim. Spektral'n. Metody Issled. Rssd Redkikh Rasseyan. Elementov, Min. Geol. Okhrany Nedr. SSSR 1961:135-139.
51. J. Owens and E.S. Gladney. 1975. Determination of beryllium in environmental materials by flameless atomic-absorption spectroscopy. Atom. Absorp. Newslett. 14(4):76-77.
52. M.A. Phillips. 1973. Levels of both airborne beryllium and beryllium in coal in the Hayden power plant near Hayden, Colorado. Environ. Lett. 5(3):183-188.
53. R.F. Abernethy and E.A. Hattman. 1970. Colorimetric determination of beryllium in coal. U.S. Bur. Mines Rep. Invest. 7452. 8 pp.
54. I.V. Vasil'eva and V.A. Vekhov. 1972. Determination of beryllium in coals and coke. Met. Koksokhim 32:100-101.
55. J. Konieczynski. 1960. Spectrographic method for determination of trace amounts of boron in coals, coke, pitch, and tars. Prace Glownego Inst. Gornictwa, Komun. (Katowice) Stalinograd Ser. B, 254, pp. 1-8.
56. V.N. Kryukova, E. Shishlyannikova, and O.I. Koralis. 1975. Boron distribution in coals of the Azeisk deposit in the Irkutsk region. Geokhimiya 7:1115-1119.
57. J. Millet. 1957. Determination of boron in industrial coals. Congr. Groupe Avance. Methodes Anal. Spectrog. Prod. Met. 10e, Paris 1957. pp. 229-235A.
58. S. Skalska and S. Held. 1956. Determination of trace contents of boron in graphite, coal, coke, and carbon black by the method of spectral emission analysis. Chem. Anal. (Warsaw) 1:294-300.
59. F.J. Allan, F.S. Dundas, and D.A. Lambie. 1969. Determination of boron content of some Scottish coals and distribution of the boron between various density fractions. J. Inst. Fuel 42(336):29-30.
60. F.H. Kunstmann and J.F. Harris. 1954. Microanalytical method for the determination of boron in coal. J. Chem. Metall. Mining Soc. S. Africa 55:12-15.
61. A.E. Vasilevskaya. 1971. Use of ternary boron compounds in analytical chemistry. Tr. Inst. Miner. Resur., Akad. Nauk. Ukr. RSR 5:22-31.
62. Y. Talmi. 1974. Determination of zinc and cadmium in

Table 10 (Continued)
Literature Available for the Determination of
Trace Elements in Coal and Coal Ash

environmentally based samples by the radiofrequency spectrometric source. Anal. Chem. 46(8):1005-1010.

63. H.J. Gluskoter and P.C. Lindahl. 1973. Cadmium: mode of occurrence in Illinois coals. Science 181:264-266.
64. K. Dittrich and G. Liesch. 1973. Atomic absorption spectroscopic determination of chromium in electrode-grade coke. Talanta 207:689-694.
65. N.P. Fedorovskaya and L.V. Miesserova. 1967. Determination of vanadium and chromium in solid fuel ash. Tr. Inst. Goryuch. Iskop., Min. Ugol. Prom. SSSR (Moscow) 23(3):3-7.
66. N.P. Fedorovskaya and L.V. Miesserova. 1967. Use of photometric method for determining the copper and nickel contents in solid fuel ash. Tr. Inst. Goryuch. Iskop., Min. Ugol. Prom. SSSR (Moscow) 23:7-15.
67. Yu. N. Pak and L.P. Starchik. 1974. Neutron-radiation determination of the iron content in coals. Koks Khim. 11:10-11.
68. N.P. Fedorovskaya, I.M. Kazakova, and N.L. Surina. 1967. Complexometric determination of Fe(III) in the analysis of ash from solid fuels. Tr. Inst. Goryuch. Iskop., Min. Ugol. Prom. SSSR (Moscow) 23(2):14-19.
69. P.J. Jackson. 1957. The determination of ferrous iron in P.F. ash and slags from pulverized fuel-fired boilers. J. Appl. Chem. (London) 7:605-610.
70. W. Radmacher and W. Schmitz. 1959. The chelatometric determination of iron and aluminum. Brennst. Chem. 40:161-164.
71. C. Block. 1975. Determination of lead in coal and coal ashes by flameless atomic absorption spectrometry. Anal. Chim. Acta 80(2):369-373.
72. F. Cuttitta and J.J. Warr, Jr. 1958. Retention of lead during oxidative ashing of selected naturally occurring carbonaceous substances. Geochim. Cosmochim. Acta 13:256-259.
73. F. Valeska and A. Havlova. 1959. Spectrographic determination of titanium and manganese in coal. Sb. Praci Hornickeho Ustavu Cesk. Akad. Ved. 1:182-190.
74. F.M. Kessler and L. Dockalova. 1955. The determination of manganese in coal ashes. Paliva 35:178-181.
75. H.L. Rook, T.E. Gills, and P.D. LaFleur. 1972. Method for determination of mercury in biological materials by neutron activation analysis. Anal. Chem. 44:1114-1117.
76. R.R. Ruch, H.J. Gluskoter, and E.J. Kennedy. 1971. Mercury content of Illinois coals. Ill. State Geol. Surv. Environ. Geol. Notes 43. 15 pp.
77. D.H. Anderson, J.H. Evans, J.J. Murphy, and W.W. White. 1971. Determination of mercury by a combustion technique using gold as a collector. Anal. Chem. 43(11):1511-1512.
78. B.W. Bailey and F.C. Lo. 1971. Automated method for determination of mercury. Anal. Chem. 43:1525-1526.
79. H. Heinrichs. 1975. Determination of mercury in water, rocks,

Table 10 (Continued)
Literature Available for the Determination of
Trace Elements in Coal and Coal Ash

coal, and petroleum with flameless atomic absorption spectrophotometry. Z. Anal. Chem. 273(3):197-201.

80. C. Huffman, Jr., R.L. Rahill, V.E. Shaw, and D.R. Norton. 1972. Determination of mercury in geologic materials by flameless atomic absorption spectrometry. U.S. Geol. Surv. Prof. Paper 880-C, pp. 203-207.

81. G.W. Kalb. 1975. Total mercury mass balance at a coal-fired steam plant. In Trace Elements in Fuel, S.P. Babu, ed. Advances in Chemistry Series No. 141, American Chemical Society, Washington, D.C., pp. 154-174.

82. T.C. Rains and O. Menis. 1972. Determination of sub-microgram amounts of mercury in standard reference materials by flameless atomic-absorption spectrometry. J. Assoc. Off. Anal. Chem. 55(6):1339-1344.

83. D. Siemer and R. Woodriff. 1974. Application of the carbon rod atomizer to the determination of mercury in the gaseous products of oxygen combustion of solid samples. Anal. Chem. 46(4):597-598.

84. M.K. Pakter. 1968. Determination of mercury in coke-chemical products. Koks Khim. 7:43-46.

85. A.E. Vasilevskaya and V.P. Shcherbakov. 1963. Determination of mercury compounds in coal. Dopovidi Akad. Nauk Ukr. RSR. pp. 1494-1496.

86. A.E. Vasilevskaya, V.P. Shcherbakov, and V.I. Klimenchuk. 1962. Determination of mercury in coal by dithizon. Zavod. Lab. 28:415.

87. K.K.S. Pillay, C.C. Thomas, Jr., and J.W. Kaminski. 1969. Neutron activation analysis of the selenium content of fossil fuels. Trans. June 1969 Ann. Mtg., American Nuclear Society, Seattle.

88. Y. Talmi and A.W. Andren. 1974. Determination of selenium in environmental samples using gas chromatography with a microwave emission spectrometric detection system. Anal. Chem. 46(14):2122-2126.

89. T.L. Belopolskaya and I.V. Serikov. 1969. Chemical and x-ray spectral determination of selenium in coals, lignites, and rocks rich in organic matter. Khim. Tverd. Topl. 6:73-79.

90. P. Buncak. 1963. Spectrographic estimation of vanadium in coke. Ropa Uhlie 5(5):144-149.

91. K. Sugawara, M. Tanaka, and A. Kozawa. 1955. A rapid colorimetric determination of vanadium in carbon materials. Bull. Chem. Soc. Japan 28:492-494.

92. V. Strnad. 1969. Complexometric determination of zinc in pitch coke. Sb. Pr. U.V.P. (Ustov Vyzk. Vyuziti Paliv) 14:56-61.

93. W. Thurauf and H. Assenmacher. 1963. Photometric zinc determination in coal and related products. Brennst. Chem. 44:21-22.

94. C. Weaver. 1967. The determination of zinc in coal. Fuel 46:407-414.

95. Zh. Ganzorig, L. Dashzeweg, O. Otgonsuren, G. Khuukhenkhuu, I. Chadraabal, and D. Chultem. 1974. Non-destructive determination of

Table 10 (Continued)
Literature Available for the Determination of
Trace Elements in Coal and Coal Ash

traces of uranium and thorium in materials of complex composition. Radiokhimiya 16(2):247-263.

96. H.G. Meyer. 1971. Non-destructive determination of uranium and thorium in geological materials by resonance neutron activation analysis. J. Radioanal. Chem. 7(1):67-69.

97. E. Donderer. 1972. Determination of hafnium, thorium and uranium by x-ray fluorescence analysis with the use of the addition method. Neues Jahrb. Mineral. Monatsh. 4:167-175.

98. D.C. Perricos and E.P. Belkas. 1969. Determination of uranium in uraniferous coal. Talanta 16(6):745-748.

99. J.N. Weaver. 1974. Rapid, instrumental neutron activation analysis for the determination of uranium in environmental matrices. Anal. Chem. 46:1292-1294.

100. N.S. Guttag and F.S. Grimaldi. 1954. Fluorimetric determination of uranium in shales, lignites, and monazites after alkali carbonate separation. Part 15. Collected Papers on Methods of Analysis for Uranium and Thorium. U.S. Geol. Surv. Bull. 1006, pp. 111-119.

101. M. Inagaki and U. Tokuga. 1959. Uranium in columns of UBE coal. Tanken 10:97-107.

102. J. Korkisch, A. Farag, and F. Hecht. 1958. A new method for the determination of uranium in phosphates, coal ash, bauxite by means of ion-exchange. Z. Anal. Chem. 161:92-100.

103. B.C. Purkayastha and S.R. Saha. 1962. On the study of rare elements in minerals from Indian sources. Part I. On the estimation of submicro amount of uranium in Indian coal. Indian J. Appl. Chem. 25(1):20-23.

104. T. Schonfeld, M. Elgarhy, C. Friedmann, and J. Veselsky. 1960. Discussion and problems of fluorimetric determination of uranium. Mikrochim. Acta 56:883-897.

105. J. Szonntagh, L. Farady, and A. Janosi. 1955. Chromatographic determination of uranium in the ash of Hungarian coals. Magy. Kem. Foly. 61:312-314.

106. C. Ujhelyi. 1955. Determination of small amounts of uranium in coal. Magy. Kem. Foly. 61(12):437-442.

107. S. Amiel. 1962. Analytical applications of delayed neutron emission in fissionable elements. Anal. Chem. 34:1683-1692.

their detection limits. In practice, it is highly recommended to employ two or more methods when possible to check accuracy. Multiple-method approaches also allow the determination of many more elements.

One important aspect to be considered is that volatilization losses of certain elements occur during certain pretreatment and ashing procedures. This can be a limiting factor on what type of sample and analytical procedure may be applied. For example, instrumental neutron activation analysis can be applied to any matrix and requires no sample pretreatment. In contrast, analysis by atomic absorption or optical emission usually requires an "ash" sample, sometimes needing extensive pretreatment. For some elements the ash data cannot be related to the whole coal. Some of these problems have been evaluated for several elements (Guidoboni, 1973; Lee et al., 1977; Pollock, 1975; Ruch et al., 1974; Schultz et al., 1975; Vasil'eva and Vekhov, 1971; Vogel et al., 1976; Weaver, 1967); however, more effort is needed.

Data on radioactivity in coal and coal-related materials are very limited. Low-level alpha-, beta-, and gamma-counting techniques have been used to assay radioactivity levels for uranium, thorium, potassium, and radium for various coals and lignites (Bayliss and Whaite, 1966; Heugi and Jedwab, 1966; Safonov, 1974; Sklyarenko et al., 1964). Elemental concentrations of uranium and thorium in coal and related materials have been determined by several methods, as listed in Table 10. The "before and after" contents of uranium and thorium in coal and ash from a coal-burning power plant have been determined (Kirchner et al., 1974), and the radium content of ~0.6 to 4.7 pCi/g of ash in flue ash has been reported (Marquardt et al., 1970).

Standards

The current coal standard most used for trace-element analysis is the National Bureau of Standards Standard Reference Material (NBS SRM) 1632. An excellent standard for determining trace elements in coal ash is NBS SRM 1633. These standards, initially circulated to many laboratories by the NBS and the Environmental Protection Agency (EPA) as an extensive comparative analysis study, were further analyzed by several activation analysis laboratories, and comparative data for 33 elements are currently available (Ondov et al., 1975). Recently, NBS SRM 1632A, a replacement for 1632, and 1635 (a subbituminous coal) have been certified for various elements. Table 11 gives the certified values for these materials.

Other comprehensive round-robin studies of coal samples, which are continuing long-term studies that involve the consideration of additional elements each year, are sponsored by the American Society for Testing and Materials (ASTM) D05 Committee on Coal and Coke. The International Standards Organization (ISO) TC 27 Committee on Solid Mineral Fuels also has extensive consensus data available for mercury, beryllium, cadmium, cobalt, fluorine, chromium, manganese, molybdenum, nickel, lead, strontium, and zinc.

In addition to consensus data, several laboratories have determined a great number of trace elements in extensive surveys of U.S. coals (Swanson et al., 1976; Carter, 1976; Gluskoter et al., 1977). Although these data, in most instances, are not confirmed, these samples

TABLE 11 Trace-Element Certified Values in NBS SRM 1632, 1632A, 1633, and 1635[a,b]

Element	1632 (Bituminous Coal)	1632A (Bituminous Coal)	1633 (Coal Fly Ash)	1635 (Subbituminous Coal)
Al	—	(3.07%)	—	(0.32%)
As	5.9	9.3	61	0.42
Sb	—	(0.58)	—	(0.14)
Be	(1.5)	—	(12)	—
Cd	0.19	0.17	1.45	0.03
Ce	—	(30)	—	(3.6)
Cr	20.2	34.4	131	2.5
Co	(6)	(6.8)	(38)	(0.65)
Cs	—	(2.4)	—	—
Cu	18	16.5	128	3.6
Eu	—	(0.54)	—	(0.064)
Fe	8,700	11,100	—	2,390
Ga	—	(8.49)	—	(1.05)
Pb	30	12.4	70	1.9
Mn	40	28	493	21.4
Hg	0.12	0.13	0.14	—
Hf	—	(1.6)	—	(0.29)
Ni	15	19.4	98	1.74
K	—	—	(1.72%)	—
Rb	—	(31)	(112)	—
Sc	—	(6.3)	—	(0.63)
Se	2.9	2.6	9.4	0.9
Si	(3.2%)	—	—	—
Ag	(≤0.1)	—	—	—
Sr	—	—	(1,380)	—
S	—	(1.64%)	—	(0.33%)
Te	(≤0.1)	—	—	—
Tl	0.59	—	(4)	--
Th	(3.0)	4.5	24	0.62
Ti	(800)	(1,750)	—	(200)
U	1.4	1.28	11.6	0.24
V	35	44	214	5.2
Zn	37	28	210	4.7

[a]Source: Ondov *et al.* (1975).

[b]All values are in ppm unless otherwise indicated. Values in parentheses are informational values only.

represent an excellent stockpile for potential reference materials. There is a need for more standards representing ranks of coals in addition to bituminous and subbituminous coals. Because some trace elements are in different chemical associations in subbituminous or lignite coals, as compared with their association in bituminous coals, the chemical analysis problems might differ significantly.

Sampling Considerations

It is mandatory that proper sampling procedures be used to obtain a representative sample and, ultimately, the desired trace-element evaluation of a particular source of coal. An excellent guideline addressing this problem has been published by the U.S. Geological Survey (USGS) (Swanson and Huffman, 1976). A time-honored procedure, published by the U.S. Bureau of Mines (Holmes, 1918), is still used by many organizations as a guide to proper sampling of coal seams. Several standards on sampling are also published by the American Society for Testing and Materials (ASTM D2234-76 and ASTM D2013-72).

Some general problems and criteria to be considered seriously for the proper sampling of coal or any heterogeneous materials have been discussed in recent symposia and in the literature (American Society for Testing and Materials, 1973; Anders, 1977; LaFleur, 1976; Maienthal and Becker, 1976).

Needs for General Analyses

There is an ongoing need for further consensus and standardization of commonly available analytical methods so that trace-element data may be more comparable from one laboratory to another. Organizations such as ASTM, ISO, and the International Atomic Energy Agency (IAEA) are actively involved in this effort; however, it is a slow process.

Besides the well-characterized NBS coal and coal-ash standards, there should be more standards avaiable for other ranks of coal.

Increasingly needed are the availability of suitable standards and convenient analytical capabilities for trace-element analysis of coal-conversion products, intermediate materials (such as tars and chars), and residues. These materials pose special analytical problems, including sampling, pretreatment, and ashing. A general concerted "stockpiling" of these coal-conversion materials is urged for future round-robin studies involving analyses of samples distributed to several different laboratories.

Although significant effort in trace-element characterization of coal-conversion systems is currently under way (Attari, 1973; Dreher, 1978; Filby _et al_., 1976; Forney _et al_., 1974; Sather _et al_., 1975; Schultz _et al_., 1977), much more is needed. This subject was discussed in terms of general areas of research needs by Magee _et al_. (1973). Several recent conferences and workshops (Ayer, 1974, 1976; U.S. Energy Research and Development Administration, 1976) discussed trace-element analysis and the need for proper sampling procedures in conversion processes in order to ensure accurate mass balances.

Further research is needed on the elucidation of specific association of the mineral state of inorganically combined elements and also on the nature of the organically combined elements in coal and related materials. These questions have been addressed by several investigators (Given, 1974; Gluskoter _et al_., 1976; Koppenaal and Manahan, 1976; Zobovic, 1976).

Analysis of Related Materials for Environmental Assessment

This vast subject covers the trace-element analysis of a whole realm of related materials, such as water, coal-processing waste streams, soils, sediments, sludge, air, and biota. The analytical methodology for these materials is extensive and can be referred to only in general terms here.

The great number of published analytical procedures for these materials are included in Analytical Abstracts, Chemical Abstracts, Nuclear Science Abstracts, INIS Atom Index, and similar journals. Extensive status reports and reviews on specific topics, such as water analysis and food, appear biannually in Analytical Chemistry (American Chemical Society, 1977). Similarly, reviews on a specific chemical procedure, such as nucleonics and ion-selective electrodes, appear biannually (American Chemical Society, 1978) in the same journal.

A series of excellent NBS-sponsored conferences over the years has been concerned with the general status of sampling, sample handling, methodology, accuracy, available standards, significance of data, and general needs of most of these specific subjects (Kirchhoff, 1977; LaFleur, 1976; Meinke and Taylor, 1972). A recent NBS publication surveys the current literature on sampling, handling, and storage of various environmental samples (Maienthal and Becker, 1976). Similar types of standards are constantly reviewed, updated, and published by the ASTM. An ACS publication specifically addresses trace elements in the environment (Kothny, 1973). The EPA has published a compilation of recommended procedures for determining various trace elements in water and wastes (U.S. Environmental Protection Agency, 1974).

The NBS, International Atomic Energy Agency (IAEA), USGS, ASTM, and other laboratories are making an effort to provide trace-element standards and consensus-data samples for a variety of environmental materials. These materials include water, air-filter deposition samples, a variety of biological materials, sediments, soils, plant materials, radioactivity standards, and rock (Flanagan, 1976; National Bureau of Standards, 1975). There is a need to fill out and to obtain consensus data for all elements under consideration in these currently available environmental standards. Although extensive data are available for some materials, other materials have data for only a few elements. One definite need is for an analytical standard as well as reliable and efficient methods for the analysis of stack gas.

GENERAL HEALTH CONSIDERATIONS

The possible significance to human health of trace metals mobilized during the coal fuel cycle is both obvious and obscure. Obvious because many metals are known to be highly toxic at high doses, but at the same time, complex pathways through the environment to man and the damage to health from low-level doses of multiple pollutants obscure the problem so much that it is difficult to draw conclusions about the real hazards. Pathways of exposure may be as direct as coal dust inhaled by underground coal miners or as circuitous as ingestion via bioaccumulation through the food chain. Fossil-fuel-cycle activities may mobilize metals over short distances, as in windblown coal fines lost during transportation, or over thousands of miles, as in increased concentrations of fossil-fuel-derived elements in the Greenland ice sheet, oceanic sediments, and midoceanic trade winds and at the South Pole (Dickson, 1972; Duce *et al.*, 1975; Weiss *et al.*, 1971; Zoller *et al.*, 1974). Trace elements may injure directly, through toxic metabolic interactions or by catalyzing the effects of other pollutants (Amdur, 1976; Nordberg, 1974). Finally, response is partially governed by individual nutritional and immunologic status (Nordberg, 1974).

Except for dust and metal assays of the lungs of deceased underground coal miners, evaluation of actual metal dose resulting from a particular process of the coal-fuel cycle is not directly ascertainable for an individual. Recent modeling efforts provide a starting point for dose assessment (Baser and Morris, 1977; Klein et al., 1975; Vaughan et al., 1975). Despite the voluminous biochemical and toxicological literature on trace-metal effects, the nature and size of the human response to the chronic, extremely low levels of expected metal exposure from coal-fuel-cycle processes is, as yet, poorly understood, as are the characterizations of emissions from many processes. Because toxicity of trace elements is often related to physicochemical states, detailed emissions characterizations are an important facet of health effects assessment (Davison et al., 1974; Linton et al., 1976; Natusch and Wallace, 1974; Natusch et al., 1974).

Results of studies directly bearing on occupational and public health effects of trace elements from the coal-fuel cycle are reviewed in conjunction with discussions of various industrial activities necessary for coal mining, processing, and utilization. Steam-electric plant coal combustion is generally considered responsible for most trace-element emissions attributable to the coal-fuel cycle; water emissions are well quantified from power plants but not good enough in other parts of the fuel cycle--extraction (underground and surface), preparation, transportation (unit train, barge, truck, and slurry pipeline), and coal gasification--to allow comparisons. Emissions for 11 trace elements to air and water from extraction, preparation, transportation, and utilization for the five major coal regions of the country in 1975 are given in Appendixes A-C. These calculations may be useful in assessing possible areas of concern to public health regionally for steps that have particularly poorly characterized emissions, such as transportation, and for making interregional comparisons.

The method of computation, described in Appendix C, should be consulted before interpreting the emissions tables and appendixes. Water emissions were not considered in the summary of calculations, because the water-emissions data are too scanty for comparative purposes.

3

COAL MINING

An understanding of coal-mining processes is germane to determining the significance and redistribution of trace elements in the environment resulting from coal mining.

Current government regulations require that before any physical activity begins at the mine, plans must be developed to assure optimum mine operation, reclamation, and conservation of the environment. National and state regulations now require that the objectives of complete reclamation and conservation of the environment be a mandatory part of mine operations (Office of the Federal Register, 1979). Environmental inventories are also required so that the site-specific concerns may be alleviated in mine planning and reclamation planning. With greater development of the western coal, a greater proportion of U.S. coal production will come from surface mines. This report has therefore given more attention to the effects on trace elements in the environment from surface mining than from underground mining. A brief description of the relevant phases of mining as practiced in the coal provinces of the United States follows.

The general, net effects of surface mining are a few-hundred-meter relocation of the overburden, the earth materials above the coal, accompanied by a reduction of its density. As a result, permeation of groundwater through it occurs more readily. The major force for the redistribution of trace elements in the environment of a surface mine is thereby related to the changed physical properties of the overburden. In an underground mine, the density of the overburden is reduced by subsidence that results in greater porosity and permeation by groundwater and air, which brings about microenvironmental changes and redistributions of the contained trace elements. Similarly, in surface mines, the removal and handling of the overburden provides for the dissolution and migration of the contained trace elements.

SURFACE MINING

Surface mines in the three western coal provinces (Northern Great Plains, Rocky Mountain, and Gulf Coastal Plain) are generally set in a rangeland environment. To a lesser extent, potential sites may occur on high, sloping plateau areas and even less frequently, on agricultural land. In the two Appalachian provinces (North and South), the environmental setting is typically hilly to mountainous. In the remaining two provinces (Eastern and Western Interior), surface mines occur in mostly agricultural environments and, to a much lesser extent, in undeveloped areas. Surface mining processes are summarized below.

Coal Extraction

The first stage of surface mining is removal of the native soil referred to as a natural soil in 91 Stat. 487 (7) (U.S. Congress, 1977). Trees and large shrubs are removed first; those of sufficient size are used as timber. Smaller vegetation is removed by dragging a chain between two bulldozers. This vegetation becomes a solid waste and is buried in the spoil area of a mine or burned on site. Soil, known as the "A horizon" in Public Law 95-87, 91 Stat. 487 (U.S. Congress, 1977) or the "natural earth materials at or vertically adjacent to the land surface with physical and chemical characteristics suitable for support of vegetation" in Title 30 CFR 211.2 (Office of the Federal Register, 1978) is removed next. This material is segregated for replacement during reclamation. The B horizon, or a combination of it and other, lower strata, is next removed and segregated for replacement as root zones. If either material must be stored, it is protected from wind and water erosion and noxious plant species in Title 30 CFR 211.40(a) (4) (Office of the Federal Register, 1978; U.S. Congress, 1977). Normal practice is to use self-loading pan scrapers (Brown, 1977). If bulldozers and front-end loaders are used, trucks transport these soils to active reclamation areas or to storage.

The next step is overburden removal. In mine areas where the overburden rock is dense and consolidated, a pattern of holes is drilled nearly to the depth of the coal. These holes are charged with explosives, which are detonated to fracture the overburden or actually move it into an adjacent, open pit. Various combinations of ammonium nitrate and fuel oil are the most commonly used explosives. Other blasting agents such as aluminum slurry mixes, gelatins, and dynamites may be used in fracturing harder formations. Following this blasting operation, overburden can be removed.

In the coal provinces of the Northern Great Plains, Rocky Mountains, Gulf Coastal Plain, and the Eastern and Western Interior, many economically significant coal beds are nearly horizontal beneath level to gently rolling surface terrain. Mining normally begins near where the coal outcrops at the surface, where the overburden is thin, then continues into thicker overburden until physical or economic limits of mining are reached. Overburden from each cut is placed in the previous cut after removal of the coal. Operation in this manner advances in much the same way as in plowing a farm field (Cassidy, 1973, p. 386). A typical section view for this operation appears in Figure 7, which shows a horizontal section of a mine plan to mine the Wadge coal seam, the uppermost coal seam (about 8 ft thick) that occurs in some portions of Routt County, Colorado (L. G. Shearer, 1975, Peabody Coal Company, Denver, Colorado, personal communication).

The upper portion of Figure 7 shows a section of a surface-mine pit looking from the highwall toward the regraded spoil. At the left end of

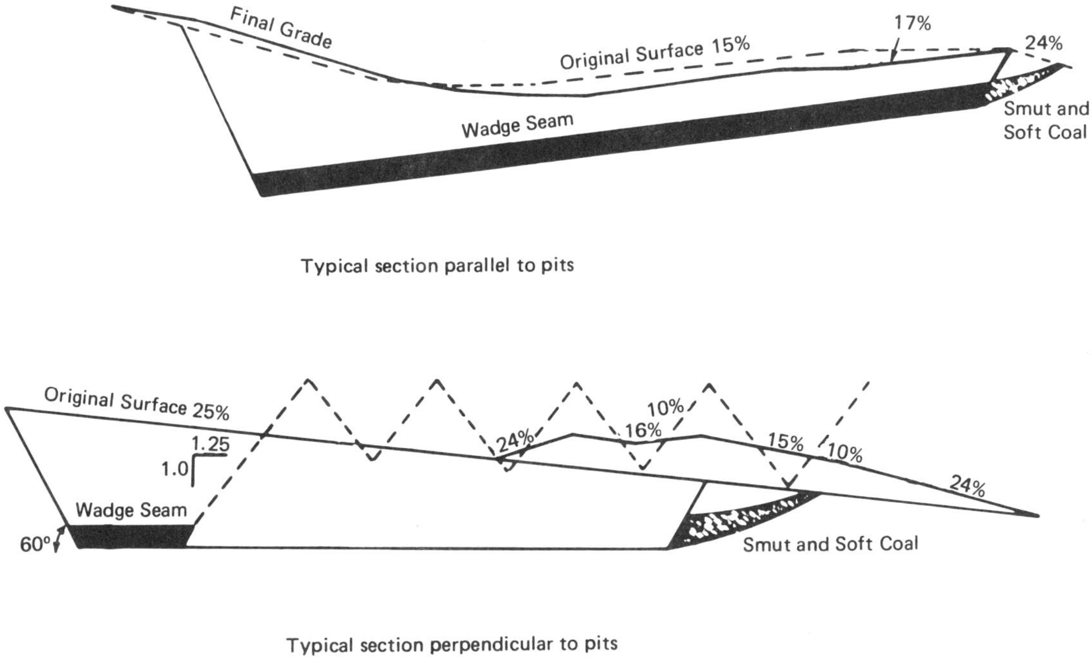

FIGURE 7 Typical cross sections of a surface-mine operation in Routt County, Colorado. (Source: Peabody Coal Company.)

the cut the depth of the coal eventually reaches a level beyond which it is uneconomical to dig. The lower portion of Figure 7 is perpendicular to the cut. The difference between the dashed line (representing the spoil as overburden is removed from above the coal and cast into empty cuts where coal has already been extracted) and the upper solid line (representing the original surface) indicates the large amount of grading that is required to meet the standards for restoration in Part 816.102 of the 1977 Surface Mining Control and Reclamation Act (Office of the Federal Register, 1978; U.S. Congress, 1977).

Figure 8 is an artist's concept of the overall operation of a surface mine. Portions of a railroad loop and coal loading layout appear near the upper left (1). Area (2) is where topsoil has been removed and set aside for later replacement on the regraded spoil. Area (3) is a bench from which the dragline will cast the underlying overburden into the adjacent cut where coal has been removed. Haul roads (4), connecting inclines (5), and topsoil storage areas (6) are shown. A fully reclaimed area (7), where vegetation has already been re-established, is now self-sustaining. A somewhat rougher area in the regrading reclamation plan (8) has been left deliberately to serve as wildlife habitat and cover.

Figure 9 is an aerial view of a coal mine that is planned and operated according to the scheme depicted in Figures 7 and 8. At the extreme left of Figure 9 is an unmined area with vegetation, a few large rock formations (which will be avoided in future mine progress), gently rolling topography, and a natural drainage channel near the top of the photo. Running vertically through the center of the photo is a strip where topsoil has been removed. In the center of the photo, a dragline

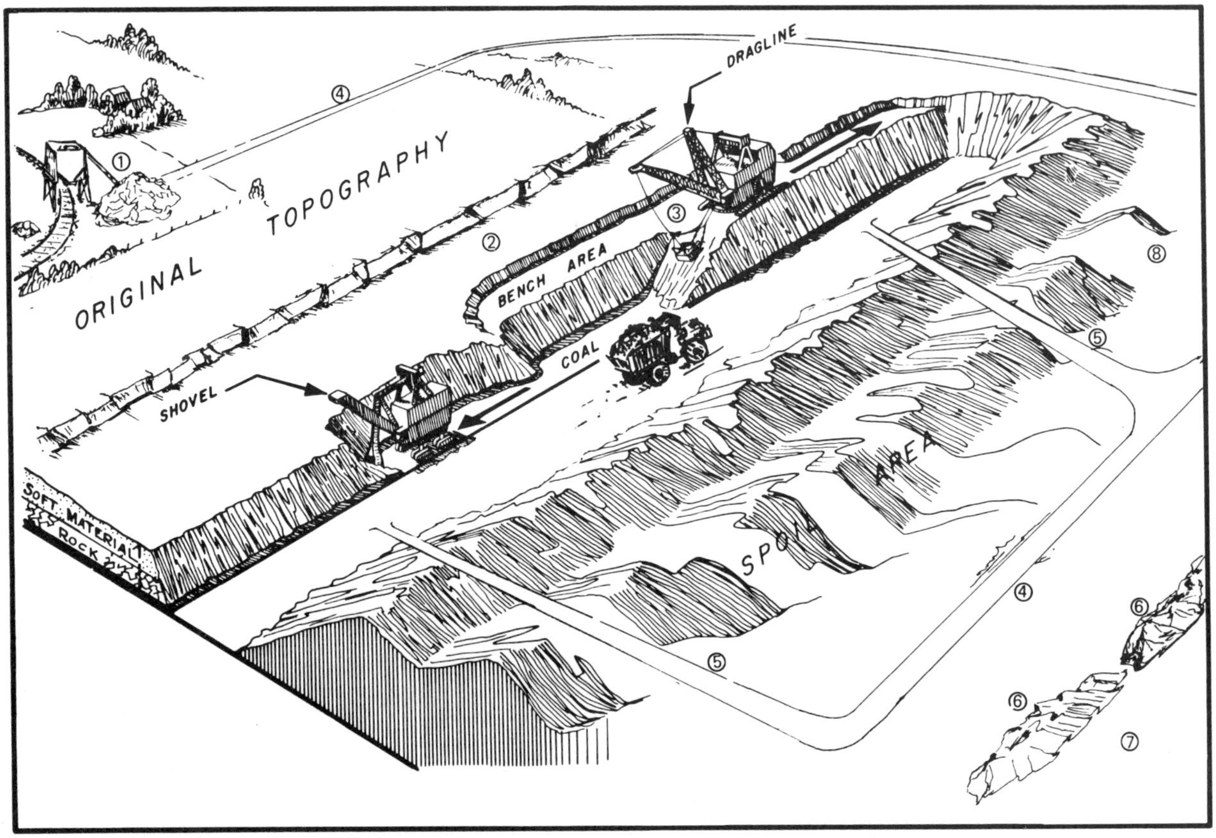

FIGURE 8 Artist's concept of a surface-mine operation. (Source: After P. Schirmer, du Quoin, Illinois.)

can be seen removing interburden (earth material lying between two coal seams) and casting it toward the right. Actual coal extraction is occurring immediately below the dragline in the darkest strip in the photo. Partially graded spoil appears in a vertical line to the right of the dragline. Completely graded spoil occurs on the right third of the photo. Inclines run across the graded spoil into the cut. This area is ready for reseeding at the next available planting season. At the extreme right of the photo, some fully reclaimed areas are partially visible (top right). Figure 10 shows a dragline in operation on an interburden bench; it is casting spoil to its right.

In other coal provinces or where the terrain is hilly, contour mining prevails. Contour mining commences where the coal bed and surface elevations coincide and proceeds along the outcrop of the coal bed. Overburden from the first cut is commonly hauled to a nearby mine pit. A second cut, and any subsequent cuts, may be made and overburden placed in earlier cuts as described above. In such instances, the cuts parallel the contact between the coal bed and the surface of the ground, unlike the plowing procedure illustrated in Figure 7.

Overburden removal by dragline or shovel mandates transfer into adjacent, open cuts. Use of bucket-wheel excavators, or other equipment using conveyor handling of dug overburden, permits moving the spoil material considerable distances from the active cut. If such an

FIGURE 9 Aerial view of a surface coal mine.

excavator is used in conjunction with a dragline or shovel, material removed by the bucket-wheel excavator will be placed in one or two cuts beyond the active pit. Use of scrapers, bulldozers, front-end loaders, haul trucks, or other such highly mobile equipment to remove overburden yields the greatest flexibility of moving spoil. With this equipment, maximum control over spoil restratification can be achieved, that is, specific layers within the overburden can be placed at particular depths within the spoil.

Most commonly, claystone, which can vary considerably in physical properties such as permeability, expansive character, and refractory quality, underlies coal beds. Layers of carbonaceous shale and siltstone commonly lie immediately above the coal bed. After coal extraction, the claystone underlying the coal bed is, of course, in direct contact with spoiled overburden. However, strata immediately above the coal bed commonly end up near or on top of the spoiled overburden.

The environmental consequence of overburden removal, coal extraction, and overburden replacement is that the lithology of the mine area changes. There is, therefore, the resulting possibility of new geochemical conditions from the claystone upward to the surface. As groundwater returns to this zone, the possibility of different chemical

FIGURE 10 Dragline on an interburden bench casting spoil to its right.

potentials exists, which may cause dissolution and migration of trace elements in this zone. Migration of and altered chemical forms of trace elements in this zone are not believed to be consequential, but this fact suggests the need for further research on the environmental impacts of mining. Investigations of this impact will be conducted as a result of the environmental regulations mentioned earlier (U.S. Congress, 1977).

In all coal provinces, most coal beds are drilled and blasted to fracture them for easier loading. The practices and explosives used are similar to those described for overburden removal. Coal extraction in a surface mine is almost exclusively accomplished by small shovels (bucket capacity up to about 30 m^3). The universal practice is to haul coal away from the active pit in trucks.

Drainage Water Control

Rain falling upslope of a surface mining area will be diverted by ditches around the active mining areas but will rejoin the natural, downslope drainage pattern. Standard practice is to design the ditches to handle a 50-year flood (the maximum flood expected in a 50-year period). Streams that normally flow at <5 ft^3/s (<0.14 m^3/s) will also be diverted around the mining area. Streams that normally flow at greater volumetric rates are usually left undisturbed and the mine designed around them.

Runoff water resulting from precipitation directly onto the active mining area joins groundwater seeping into the mine. Groundwater seepage occurs from the highwall, the pit ends, and, to a lesser extent, the spoil side of the pit. Such water is collected in sumps and pumped to settling ponds, where it is treated as necessary to conform to the National Pollutant Discharge Elimination System (NPDES) [Federal Water Pollution Control Act, Amendments of 1972, Public Law 92-500 (U.S. Congress, 1972)] prior to release to the natural downslope drainage pattern.

UNDERGROUND MINING

Coal Extraction

In an underground mine, some digging operations are required, such as shaft construction, that produce rock but not coal. This ordinary mine rock generally must be brought to and handled on the surface. Further, some roof and floor rocks are extracted with the coal during mining underground to assure maximum coal recovery. A crushing, grinding, and screening operation at the mine separates most of the roof-and-floor rock from the coal. This rock is generally disposed of with the other rock produced at the mine, commonly to piles on or near the mine site, or it is used for roadfill and building low dams for water impoundments. The volume of rock thus affected, with its trace-element loading, is small in comparison with the volume of rock similarly disturbed in a surface-mine operation. This waste rock, however, may contain some waste coal. If it does, such waste is heaped into piles, and drainage from upslope is diverted around them. They are covered with earth and the sides of the piles may be further sealed with clay. Covering and sealing prevent spontaneous combustion of waste coal that may be present. Rolling and compacting the ends and sides reduce erosion and slumping, which might otherwise expose waste coal.

Subsidence

Subsidence commonly occurs following coal extraction by underground mining. In a few places, subsidence occurs relatively soon, as a result of controlled roof collapse, during long-wall mining. In other places, as in some old mining areas, subsidences may continue for several decades after mining has ceased. In some mines, the floors of underground mines heave upward, thereby reducing the amount of subsidence, but subsidence, in some degree, occurs above practically all underground coal mines.

As controlled roof caving occurs, subsiding rock fills the void left from coal extraction with less dense material. Subsequently, additional overlying material, but with less volume, settles into the void left in collapse of the immediate roof. This sequence repeats. As the effect propagates to the surface, broken rock gradually fills the void.

This generally accepted picture of how subsidence occurs produces at least one significant result with respect to the impact on trace elements from coal extraction from an underground mine, that is, the reduced density of the geologic zone immediately above the extracted coal. The reduced density results in greater porosity and therefore greater surface area of solid exposed to groundwater. The consequence is a greater potential for dissolution and then migration of trace elements in the immediate environment of the mine. As in the case of surface mining discussed above, this phenomenon bears further investigation to determine its quantitative effects on trace elements as a result of coal extraction. These investigations will be conducted in the future because of environmental factors reporting required by current regulations governing coal mining.

Groundwater Control

Dewatering of underground mines requires collection of mine water in a sump and pumping it to settling basins (which now commonly include water treatment), either outside the mine or in abandoned workings. The pumping of mine water to mined-out areas is possible in relatively dry mines, and such action may serve eventually to recharge aquifers. In new mines with water passing through undisturbed earth, the effects of this operation on the redistribution of trace elements may be no more significant than from a well in an undisturbed area. On the other hand, if seepage eventually occurs from old workings to surface streams, the effect is about the same as the pumping of such water to surface settlement basins, with the added leachates from any earth filling in the old works. The effects of this latter process on trace-element redistribution appear similar to those of subsidence.

Acid Mine Drainage

Trace elements from acid mine drainage (AMD) may be mobilized into subsurface water systems, lakes, and streams. In the late 1960's, AMD was especially serious in the eastern half of the country, where over 10,500 miles of waterways were being degraded by acid drainage and 200,000 acres of land became barren or infertile because of refuse disposal and acid runoff (Ohio Basin Regional Federal Water Pollution Control Administration, 1969). By 1977, these figures were reduced to 5700 miles (nearly half of the waterway mileage) and to only tens of thousands of acres of land (Kendrick, 1977).

Today, AMD is primarily effluent from abandoned underground mines. A survey in Appalachia indicated that inactive underground mines contributed 52 percent of the total acid drainage in the region; active underground mines contributed 19 percent. The acid drainage may continue to flow from inactive mines as long as air, water, and sulfide materials are available. The areas most seriously affected are those where mining has been extensive and where pyritic materials are abundantly exposed to oxidation. Western coals generally contain small amounts of pyritic materials, so wastes are usually nonacidic (Ohio Basin Regional Federal Water Pollution Control Administration, 1969; Wewerka et al., 1976b). The burgeoning use of coal is not expected to reverse this favorable trend in controlling AMD because current regulations applicable nationally do not allow degradation of existing, receiving-water quality. The Surface Mining Control and Reclamation Act of 1977 on 91 Stat. 488 (U.S. Congress, 1977) specifically prohibits discharge of water from a mine that does not meet local stream water-quality standards. Thus, governmental regulations have greatly reduced the possibility of increased AMD.

The iron sulfide minerals, pyrite and marcasite, exposed to air produce ferrous sulfate and sulfuric acid. The oxidation of one molecular weight of pyrite (FeS_2) ultimately leads to the release of two molecular weights of sulfuric acid. Secondary reactions of sulfuric acid with minerals and organic compounds produce concentrations of a wide variety of elements in the drainage that may persist long after acidity has been neutralized.

The AMD control methods fall into two categories: those directed at preventing the formation of AMD and those that treat AMD after it has formed. Although these methods are usually concerned more with reducing acidity than with reducing metal concentrations, concentrations of many metals are correspondingly reduced as the solution pH is raised above neutrality.

Techniques for controlling air and water influx--such as compacting or sealing the wastes, grading and covering the wastes with soil, and burying the wastes in underground or strip mines--reduce the levels of dissolved or suspended trace metals in the runoff. Neutralization of wastewaters with alkaline agents, including limestone, lime, and caustic soda, is the most widespread treatment method, with some 300 plants operating in the United States. These treatments are designed primarily to treat acidity; only those ions, such as iron and aluminum, and certain trace elements whose solubilities are sensitive to pH in the acid to neutral range will be removed by the alkaline neutralization process. The solubilities of other ions, such as calcium, magnesium, sodium, potassium, and sulfate, are not highly dependent on solution pH (Wewerka et al., 1976a) and will not be reduced; they may even be increased.

Methods of treating AMD using ion exchange, reverse osmosis, flash distillation, and dissolved ions, which have been developed, are not currently used on a wide scale commercially. The sludges produced by ion-exchange methods appear to be easier to dewater than those produced by alkaline neutralization, but the methods are limited in application to drainage with medium- to low-level solids (<1000 ppm).

Surface mines may also be sources of AMD, though much less so than underground mines. Inactive surface mines contributed 11 percent of the AMD in the Appalachia inventory, and active surface mines contributed 1 percent (Ohio Basin Regional Federal Water Pollution Control Administration, 1969).

RECLAMATION

Overburden and Soil

After grading of the overburden, topsoil is replaced on the area. The topsoil may originate either from storage or from immediate salvage. The topsoil is smoothed by grading also. At this point the area is ready for revegetation at next planting season.

The first step in reclamation is the development of a plan for postmining land use. In the western coal provinces, the typical postmining land use is currently rangeland and wildlife habitat. To a lesser extent, water storage and recreational uses are intended. In other coal provinces, however, a combination of crop production, pasture and rangeland, wildlife habitat, and recreational area is typically negotiated among several community groups. The specific combination of uses dictates the overall objective of the spoil-grading plan and influences the redistribution of trace elements therein.

Reclamation of spoil to produce pasture or rangeland for grazing can be accomplished with less grading than for crop production. Commonly, ponds are included in reclamation plans based on pasture uses for domestic-animal water supplies. Topsoil is also replaced and smoothed over the graded spoil.

Reclamation to produce wildlife habitat requires rougher terrain in the spoil. These rougher areas approximate natural draws or gullies and thereby provide protection from visibility for wildlife. Location of such draws will coincide with spoil drainage channels so that water supply to vegetation is assured, further protecting wildlife by additional cover, as well as providing food.

The extent of grading required to provide recreational areas depends on the type of recreational use intended. Rougher terrain suits camping and hiking; smoother terrain must be produced for park-like areas. Large water impoundments may be included and stocked with game fish.

Postmining Spoil and Overburden Drainage

Final or initial cuts can serve the useful purpose of surface water storage. In this case, the edge of the coal seam at the base of a "highwall" is covered by impermeable sediment to prevent direct contact between the coal and the stored water. Exchange between the stored water and the coal seam is only through the groundwater regime. Where upslope runoff or spoil-area runoff is directed to these catch basins, they serve the subsidiary, useful purpose of sediment entrapment.

Grading of the spoil must create channelization of runoff to the area planned for water impoundments. Knowledge of groundwater production and movement in the area and of local precipitation patterns is necessary to plan a grading pattern for this purpose.

To maximize retention of runoff water as aquifer recharge, grading must create a series of smaller catchment basins. Catchment basins are smaller and shallower than impoundments. Wide distribution of such catchment basins encourages percolation of collected runoff into the groundwater regime. Acid- or toxic-producing materials occurring in the spoil are buried deeply to avoid contact with runoff water. Exclusion of water contact can be through burial under (or within) an impervious material such as clay in Title 30, CFR 211.40(a)(1) (Office of the Federal Register, 1978; U.S. Congress, 1976, 1977).

Refuse and Waste Disposal

Reclamation in coal mining includes disposal of refuse and waste from mining. Refuse is generated from ancillary processes, which may accompany coal extraction, such as crushing, sizing, and washing (i.e., coal preparation). In addition, office, shop, and maintenance areas generate waste.

Coal preparation produces a slurry composed of finely ground coal and inorganic minerals in the coal, such as quartz and pyrite. Typically, this slurry is pumped to settling basins where solids are retained and the clarified water is discharged if or when it conforms to NPDES regulations [Federal Water Pollution Control Act Amendments of 1972 (FWPCA) Public Law 92-500 (U.S. Congress, 1972)]. As slurry settling basins become filled with sediment, they are drained and covered with natural earth material from the surrounding terrain. Slopes on the burial areas (shown in cross-sectional view in Figure 11) are compacted with rollers.

The mining process, apart from overburden removal, coal extraction, and coal preparation, also produces wastes that must be handled. In a surface mine, these wastes result from cleaning the top of an exposed coal seam before the coal is removed. Extraneous rock and coal refuse are scraped from the top of the seam with a bulldozer and pushed into the adjacent pit; this material is then covered with spoil as the excavation machine makes its next cut to remove the overburden. This material is relocated vertically and horizontally by only a matter of meters but is broken up in the process. The included trace elements are exposed to a new microenvironment just as in the case for overburden and topsoil.

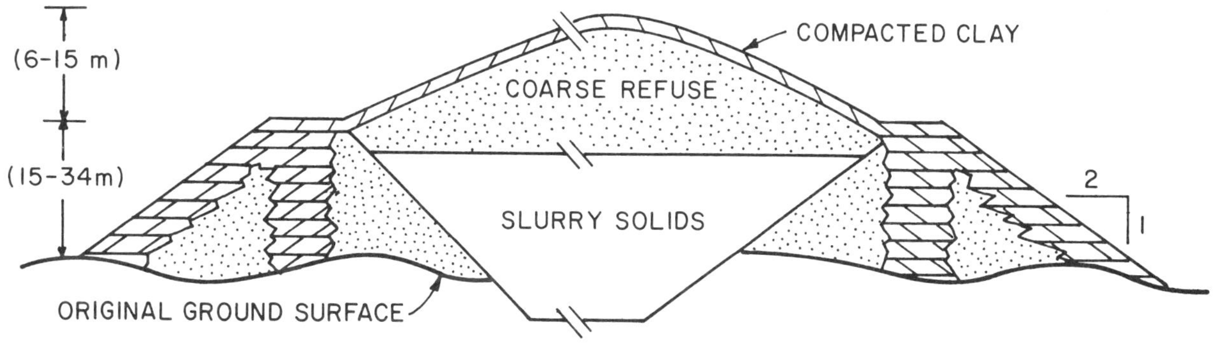

FIGURE 11 Cross section of buried slurry settling basin.

Final Abandonment

By the time of final abandonment, the area of active surface mining should be completely reclaimed to the state of a self-sustaining ecosystem. Structures are demolished or removed for reuse elsewhere. Waste materials from abandonment of facilities are disposed of in a manner similar to a sanitary landfill. Once they are buried in the sanitary landfill, revegetation proceeds by seeding, planting, fertilizing, and irrigation, as necessary. Compacted areas are scarified, soil amendments are added as necessary, and seeding re-establishes vegetative cover. Other alternative land uses may dictate continued use of certain of the access roads or facilities. In this case, continued concern for redistribution of trace elements in the environment passes from mining to new uses of such facilities.

Revegetation Processes

Some trace elements are undoubtedly redistributed as a result of steps taken to revegetate areas disturbed by mining. New or different sources and amounts of trace elements are made available to vegetation, and their potential beneficial or detrimental effects should be assessed through controlled environmental or agricultural studies on test plots on reclaimed land with careful sequential chemical analysis of product growth. Disturbed areas include not only regraded areas of actual surface mining but also areas at abandoned mines where facilities and haul roads and/or access roads are removed. Such disturbed areas also include areas at underground mines where surface use was required during operation of the mine. Dissolution of trace elements by percolating surface water or groundwater makes them available to uptake by vegetation. The significance of trace-element uptake by vegetation depends on the type of vegetation and the intended use of the vegetation. The role of the vegetation from reclaimed areas in the human food chain is a further factor influencing this significance.

Reclamation plans for mining areas do not always call for revegetation because some reclaimed areas may be used as urban or industrial sites. However, most reclaimed areas are to be used as agricultural or recreational areas.

Redistribution of trace elements as a result of revegetation on agricultural or recreational areas generally begins with the increased availability of trace elements to the new vegetation because of soil and rock disturbance. Dissolved salts of trace elements in the soil are subject to uptake by vegetation through the depth of root penetration, which will vary with different types of vegetation; rarely will such a depth be greater than 20 ft (6 m).

Where vegetation is not removed from the site through agricultural crop harvesting, further redistribution of trace elements will be limited to airborne scattering of dead and dried vegetation. This redistribution would be minor in comparison with redistribution through agricultural crop harvesting. A crop such as hay, corn, or other similar vegetation would be harvested, removed from the site with its component trace elements, and fed to livestock or people. In any case, man is at the end of this food chain, and his ingestion of the trace elements removed from the mine site is of ultimate concern.

Vegetation on the reclaimed site also may serve as a food source for wildlife, and the wildlife, in turn, may be eaten by man; for example, deer or antelope may eat the vegetation and then be consumed by man. The magnitude of this redistribution of trace elements from the mine site would probably be small in comparison with consumption of vegetation by livestock and thereby introduction into man's diet.

REDISTRIBUTION OF TRACE ELEMENTS IN MINING

Trace elements that occur in coal will be removed from their in situ location, transported to a coal preparation site, and thence to the site of coal utilization. Topsoil handling, overburden removal and replacement, coal extraction, revegetation, and subsidence may influence trace-element redistribution. Redistribution can occur through one or more of the following processes:

- Physical relocation.
- Mechanical breakup (fragmentation) of previously consolidated material that increases the surface area of rock so the rock and minerals are more exposed to weathering and subsequent possible trace-element mobility.
- Major change in porosity and permeability of rock material, with a consequent increase in the rate and amount of water that moves through near-surface aquifers.
- Change from chemically reducing conditions to oxidizing conditions, which alter the solubility by conversion to the oxidized forms of the trace elements.
- Oxidation of pyrite and release of acid, thereby enhancing solubility and mobilization of trace elements.

All of these processes may influence the availability of trace elements to the biota. Each of the mining phases produces a new microenvironment for the trace elements. The creation of this new microenvironment is the single most important factor influencing redistribution of trace elements in the mine environment. This section addresses the potential means for redistribution of trace elements in the environment as a result of mining. First, there is a simple relocation of the trace elements in situ in native soil by topsoil salvage, overburden removal, and reclamation. Several subprocesses involved in this part of coal mining are identified. There is, in general, inadequate information available to evaluate the importance of these subprocesses individually. Additional empirical and theoretical evaluation of the contribution of each subprocess to the overall redistribution of trace elements is required. The new microenvironment results in part from juxtaposition of strata that were not so located before mining. New information about the potential contribution to redistribution from the new microenvironment is needed.

The phases of mining described cause changes in the environment of the mine; these mobilization processes can redistribute trace elements.

Topsoil Handling

The replacement of topsoil from salvage site or active mining site to graded spoil is, in itself, a redistribution of intrinsic concentrations of trace elements. The topsoil may be picked to remove stones above a certain diameter, say 15 cm. Picking of stones and their burial in the mine pit are a secondary-level redistribution of intrinsic concentrations of trace elements. The degree of consolidation, or density, of replaced topsoil will be less than that of in-place topsoil. Further, the chemical and physical nature of subsoil on which the topsoil is placed may be different from that at its original site. Because of the slope, the exposure to the sun may also be different. Finally, the moisture conditions may be changed in the topsoil as a result of both surface water and groundwater changes.

Salvaged topsoil that is placed in storage can be leached of some trace elements. The factors that affect leaching from stored topsoil include:

- geometry of the storage pile,
- amount of precipitation,
- rate of precipitation,
- chemical nature of precipitation,
- availability of groundwater below the pile,
- drainage away from the pile, and
- length of time in storage.

The geometry of a topsoil storage pile affects absorbency and the percolation of the precipitation falling on it. Conically shaped piles reduce absorbency and thereby percolation of runoff water into the pile. Where the geometry of the pile presents more surface area to runoff water, it follows that absorbency would be increased.

Trace elements can also be transferred from topsoil storage during wind erosion associated with dust blown away from the stored topsoil. Normal practice in the mining industry now is to control such dust by spraying the surface of storage areas with water until temporary vegetation can be established to hold the surface soil in place.

Overburden Removal and Replacement

At most surface mines, spoil piles are graded to a gently rolling contour. The grading operation relocates trace elements occurring in the original overburden. This relocation is a minor effect, as is overburden removal, compared to the potential and actual redistribution of trace elements by oxidation and water-solubilization processes occurring at the microenvironmental levels. In any method of overburden removal, the density of the spoil will be less than that in the overburden. The general rule is that overburden will swell by approximately 20 percent. Thus, as the density decreases with swell, the surface area per unit of volume of the overburden, or spoil, increases. As the surface area increases, the potential for oxidation and redistribution of trace elements from the spoil also increases.

The decreased density in spoil may change the mode by which water moves through the space originally occupied by overburden. Water moves more readily by percolation through the greater porosity (resulting from decreased density) of the spoil than through in situ overburden. More particle surfaces of spoil are in direct contact with the percolating water, hence the potential for redistribution of trace elements is further enhanced. Actual redistribution of trace elements will depend not only on the chemical nature of the overburden but also on the chemical nature of the groundwater. The principal mechanism for redistribution of trace elements in a mining environment is thought to be dissolution of their salts from the spoil, followed by migration of the trace elements via groundwater. The primary phenomena that cause redistribution of trace elements via groundwater are the solubility of the trace-element salts and their movement with the groundwater (Bird et al., 1960). The groundwater may seep out of the area and add to downslope drainage and flow. Such downslope drainage might eventually contribute to drinking-water supplies for either man or livestock. In either case, the trace elements carried from the mine site by groundwater seepage eventually may be consumed by man, livestock, or wildlife as drinking water.

Dissolution of trace-element salts in groundwater permeating the spoil depends on the chemical nature of the groundwater. The definition of this relationship through an application of well-known thermodynamic and kinetic principles and data in available reports might assist the mining industry in predicting the concentration of trace-element salts in groundwater seeping out of the mine area. The interdependence of overburden characteristics, the availability and composition of groundwater as a transport medium, and the method of overburden removal and replacement make the analysis of potential redistribution of trace elements a site-specific problem.

HEALTH EFFECTS

Occupational Health

Inhalation of coal dust and concomitant trace-metal exposure from surface mining have not been shown to be hazardous. The U.S. Public Health Service surveyed 1438 surface coal miners to determine the prevalence of Coal Workers' Pneumoconiosis (CWP), chronic bronchitis, and ventilatory impairment. Four percent of miners showed some x-ray evidence of CWP, but all had previously worked in underground mines over prolonged periods. The study concluded that employment in surface mines was not likely to cause CWP or clinically significant respiratory impairment (Fairman et al., 1977). There is, however, some current concern regarding possible detrimental effects from exposure to exhaust fumes and noise generated by the heavy equipment used in surface coal mining. Possible effects from these exposures are now under study by several federally funded projects.

Coal Workers' Pneumoconiosis is a well-recognized hazard of underground coal mining. The multiple respiratory disorders associated with CWP vary in prevalence and severity according to geographic area, degree of occupational exposure, and apparent individual susceptibility

(Naeye, 1972). Study of 1455 bituminous coal workers from six mines revealed a 47 percent prevalence of x-ray evidence of CWP and a 2.4 percent prevalence of progressive massive fibrosis (PMF), as compared with 60 percent and 14 percent prevalence rates of CWP and PMF, respectively, for 518 anthracite miners from two anthracite mines (Morgan et al., 1972).

The National Coal Board followed 4122 coal mine-face workers at 20 collieries, relating total dust concentration and mineral content of dust to x-ray evidence of CWP and progression to PMF and to lung pathology at autopsy (Davis et al., 1977; Walton et al., 1977). The amount of total dust exposure was the most significant factor in determining probability of progression to PMF. Mineral content was a lesser factor, but the probability of progression appeared to fall with increasing mineral content. Mineral content in this study was predominantly quartz. The authors suggested that the apparent fall in toxicity of the dust with increasing mineral content might alternatively be explained by a diminishing effect of coal of decreasing rank (Walton et al., 1977).

Several CWP-related disorders lead to pulmonary fibrosis. In some cases the fibrosis is clearly related to mine dust; in others the pathogenesis is unclear. There is extensive literature documenting the toxicity and fibrogenecity of free silica on cells in culture, and the silica content of miners' lungs is directly related to the collagen content of the macules and nodules (Naeye, 1972; Wagner, 1972).

Certain trace elements could act as factors or cofactors in fibrogenesis. A long-range study of lung tissue from deceased bituminous coal miners from West Virginia compared metal concentrations between miners and nonminers and examined the relationship of total dust, free silica, and trace-metal concentrations to severity of CWP (Sweet et al., 1974). The concentrations of iron, nickel, titanium, and vanadium in miners' lungs were markedly higher than in lungs of long-term residents (nonminers) of the same county as the miners. The miner:nonminer concentration ratios were 2:1 for iron, 4:1 for nickel, 4.5:1 for titanium, and 4:1 for vanadium. A statistical test for trends in classification means (Duncan's multiple range test) indicated a continuously increasing trend in the means of magnesium, beryllium, vanadium, free silica, and coal dust with increasing degrees of lung damage. There was no relationship between the severity of CWP and means of chromium, copper, iron, manganese, nickel, titanium, zinc, and noncoal dust. An analysis of variance indicated that vanadium was more strongly associated with CWP than either beryllium or magnesium. Compositional differences have been found between the characteristic areas of heavy and light pigmentation of pneumoconiotic lungs. Bismuth, magnesium, manganese, and germanium were more concentrated in the areas of heavily pigmented tissue from two pneumoconiotic lungs than in the lightly pigmented tissue (Sweet et al., 1978).

An earlier report showed that only silica and vanadium, but not coal dust, total dust, iron, nickel, lead, zinc, manganese, beryllium, titanium, copper, magnesium, or chromium, were significantly more concentrated in hilar lymph nodes than in lungs (Carlberg et al., 1971). Silica was found to be more concentrated by a factor of 3.6, and vanadium by a factor of 1.63. These findings suggest that vanadium is present in an insoluble, stable form.

Vanadium in coal is present not only in inorganic forms but also in vanadyl porphyrin chelates (stable five- and six-member rings) associated completely with the organic fraction of coal. The toxicity via the inhalation pathway of this form of vanadium is not known, although vanadium pentoxide is severely irritating to the mucous membranes of the eyes, nose, throat, and respiratory tract (Waters, 1977).

While there is clearly enrichment of certain minerals and elements in the lungs, cause and effect are difficult to determine. The differential enrichment may be a result of differential retention caused by the disease, rather than a cofactor in causing the disease (Davis et al., 1977).

It may be that incidence of CWP and PMF is determined partly by susceptibility in the miners. Heise et al. (1977) showed that the relative risk of PMF was approximately 300 percent less in miners who possessed antigen W18 than in those lacking this antigen.

The predominant radionuclide emission from coal mining is radon, released into the air during exposure of the seam and breakup of the coal. A survey of the presence of ^{222}Rn and ^{220}Rn daughters in the range of 0 to 0.3 working levels in 223 operating mines in 15 states concluded that there was no apparent occupational health hazard from inhalation of the ambient levels of ^{222}Rn or ^{220}Rn daughters (Rock et al., 1975).

Public Health

Acid mine drainage (AMD) can adversely affect water supplies. Effects may include increased acidity, discoloration, and increased concentrations of iron, manganese, silt, coal fines, and sulfates. The high acidity and increased corrosiveness might interfere with water treatment processes, such as coagulation and softening, although the mineral content can usually be reduced by additional treatment. The direct public-health effects of increased metal concentrations in drinking water are unclear, especially as only a small fraction of the daily intake of elements reaches the human body through drinking water (Table 12), as the amounts reaching man via bioaccumulation through the food chain and relative uptakes from various sources have not been quantified. Surveys by the EPA Effluent Guidelines Division have confirmed the presence of several organic petroleum and solvent types of compounds in drainage water affected by mining operations. The pollutants, among which are several suspect carcinogens, are believed to originate in equipment-servicing operations.

Research on coal, trace elements, and mortality is being attempted by Kagey et al. (1978) to determine whether the identity and geographic distribution of dissolved trace elements in water can be correlated with epidemiological data.

Trace-element air emissions from surface mining result from diesel-fuel combustion and windblown coal fines. The regional emissions are low, ranging from 10^{-2} ton/yr to 10^{-6} ton/yr for various metals (Appendix C). Strip mining is most prevalent in the Northern Plains and the Southwest, but the public health risks are attenuated by the low population densities in these areas.

TABLE 12 Total Daily Intakes of Various Elements from Food and Water[a]

Element	Intake from Water		Average Intake from Food and Water (mg/day)	Proportion of Total Intake from Water	
	Median (mg/day)	Maximum (mg/day)		Median (percent)	Maximum (percent)
Essential					
Calcium	52	100	800	6.5	11.8
Magnesium	12.5	40	210	5.9	16.8
Sodium	24	100	4400	0.5	2.2
Potassium	3.2	10	3300	0.09	0.3
Vanadium	<0.008	0.02	2	0.6	1.0
Chromium	0.001	0.01	0.1	1.0	9.2
Manganese	0.01	0.2	3	0.3	6.3
Iron	0.09	0.3	15	0.6	2.0
Cobalt	0.006	0.01	0.3	2.0	3.3
Nickel	0.005	0.02	0.4	1.3	4.8
Copper	0.02	0.2	2.5	0.8	7.5
Zinc	0.5	2.1	13	3.8	14.4
Selenium	<0.02	—	0.15	<13.3	—
Fluorine	0.4	1.0	1.8	22.2	41.7
Molybdenum	0.003	0.02	0.34	0.9	5.6
Nonessential					
Silicon	14.2	60	<20	<71.0	<90.0
Aluminum	0.1	1.0	45	0.2	2.2
Barium	0.09	0.76	1.24	7.3	39.8
Strontium	0.22	1.0	2	11.0	40.0
Boron	0.06	0.2	1.0	6.0	17.5
Bismuth	Trace	—	0.002	—	—
Beryllium	Trace	—	0.00001	—	—
Antimony	Trace	—	<1.0	—	—
Lead	0.007	0.02	0.41	1.7	4.7
Lithium	0.004	0.1	2.0	0.2	4.7
Silver	0.0005	0.001	0.07	0.7	1.4
Tin	0.002	0.005	4.0	0.05	0.1
Titanium	<0.003	0.0	0.3	0.1	3.2
Uranium	0.0003	0.004	1.4	0.02	0.3
Cadmium	0.005	0.04	0.07	7.1	38.1

[a]Sources: Schroeder (1973) and Berry and Wallace (1974).

Comprehensive studies of radionuclides in coal mine drainage have not been undertaken, but some measurements have been reported (Caldwell et al., 1970). Gross alpha levels up to 180 pCi/liter were found in the Kiskiminetas River of Pennsylvania. Radioactivity was predominantly from ^{234}U and ^{238}U.

4

COAL CLEANING, STORAGE, AND TRANSPORT

Information is sparse concerning the evaluation of potential trace-element problems encountered in the cleaning, storage, or transport of coal from the mining and processing point to the ultimate consumer. The potential contribution of undesirable concentrations of trace elements to the environment as a result of the storage or transport techniques currently used for coal in the United States has been questioned (Swanson *et al*., 1974). The purpose of this section is to examine the trace elements that are known to occur in coal and their potential impacts during coal cleaning, storage, or transport.

COAL CLEANING

Coal cleaning is conventionally based on the principle that the particles of crushed coal differ from each other in the relative proportions of included mineral matter and organic combustible material. Most coal cleaning occurs at or near the mining area, often just before the coal is transported. Because mineral-rich particles have higher densities than organic-rich particles, they can be separated by processes that use a liquid medium with a specific gravity that will optimize the separation of coal from mineral matter. A detailed description of an intensive coal-cleaning process is given by Ford *et al*. (1976). However, coal cleaning does not eliminate any elements, it simply separates the mineral-associated ones from the organic fraction of the coal. The trace-element content of the cleaning water is tremendously enriched, which greatly magnifies the runoff problems beyond those encountered in mining and storage. However, none of the referenced studies contain data suitable for use in this report. Although changed in concentration and form, the disposal of mineral constituents is still necessary. The cleaning operation might dissolve some elements, which could result in a water-disposal problem. The average percent removal of arsenic, lead, manganese, mercury, and selenium during coal cleaning as calculated from data in Ford *et al*. (1976) ranged from 28 percent for selenium to 41 percent for arsenic (Table 13).

The Southern California Edison Company Mojave Generating Station (at Bullhead City, Nevada) has a unique coal-delivery system (Environmental

TABLE 13 Percentage of Trace Elements Removed by Coal Cleaning[a,b]

Element	Percentage Removed	
	Average	Range
Arsenic	41	11-67
Lead	32	8-63
Manganese	37	9-76
Mercury	30	3-68
Selenium	28	2-61

[a]Source: Ford *et al.* (1976).

[b]Coals used were bituminous from Kentucky, West Virginia, Pennsylvania, Oklahoma, Kansas, and Alabama.

Science and Technology, 1976). Coal is transported from the Black Mesa field (Arizona) as a 1:1 coal-water slurry in an 18-in. pipeline over a distance of 275 mi (440 km). On arrival, coal is centrifuged at 1000 G, and 75 percent of the water is removed before the coal is introduced into the boilers. Such a system could combine transportation with a degree of cleaning and separation of undesired constituents. However, such a separation might result in dissolution of undesired constituents and increased water-disposal problems. Samples of the coal slurry water obtained at the pipeline terminus were found to be highly acidic, with an average pH of 3.3 for six samples. The acidic condition was attributed to the presence of pyritic materials in the coal. Slurry water was also elevated in total soluble salt content, with an electrical conductivity of 3.1 mmho/cm, corresponding to approximately 2000 ppm of total dissolved solids. Calcium, sodium, and magnesium were the dominant cationic species contributing to the salinity of the slurry water (A. L. Page, University of California, Riverside, personal communication, September 1978).

Coal preparation plant wastes and reduction of sulfur and trace elements have been studied by the National Research Council (NRC) Committee on Accessory Elements' Coal Panel (1979). Because of this, further discussion on coal preparation will be limited. However, the slurries or residues remaining from coal-cleaning processes should be site-specifically analyzed and monitored to control the concentration of trace elements originating from this process.

STORAGE AND TRANSPORT

According to Gleason (1976), dust has been one of the major problems encountered with the stockpiling, loading, and unloading of western coals. Control was implemented at the Montana facilities through the use of paved roads. Watering trucks were equipped to spray unpaved haul routes and storage piles. In some instances, a surfactant was added to the water, and about once a week a crusting agent was applied to the inactive parts of the work yard or to the sides of coal storage piles to control dust and inhibit rainwater percolation.

Ehrlich (1976) also reported that the handling and storage of some western coals (mainly lignite and subbituminous coal) present a problem in that the coals are friable and may spontaneously ignite. Prevention of spontaneous combustion in these coals requires that they be either compacted, intermittently agitated, or stirred during storage in stockpiles or during rail transport and that their moisture content be carefully monitored.

The Committee on Environment and Public Planning (Swanson et al., 1974) of the Geological Society of America has noted the need to control coal dust during handling and storage to prevent dust explosion and spontaneous combustion hazards. They also recognized that trace metals may be leached or redistributed from coal storage areas, resulting in localized elevated levels of trace-metal concentrations.

Coal and other fugitive dust emissions have been studied by Amick et al. (1974). They suggest that a total emission inventory must be used to ascertain impacts, to provide baseline data, and to predict effectiveness of long-range control strategies in meeting future air-quality needs.

According to Ross (1977), elevated trace-element concentrations have been found associated with soils surrounding coal storage or other areas susceptible to blowing dust near railroad transportation areas. In this literature review concerning trace-element entry into the soil environment, such factors as aerial deposition, pH, fallout during rain, and weathering are cited as topics of concern.

More definitive trace-element research has been performed on the contaminants that originate when coal is processed (Wewerka et al., 1976a). Although the emphasis was on coal refuse dumps, the results might have application to coal stored on ground facilities or to blowing materials from rail or haul roads.

Another rather extensive review on the chemistry and behavior of trace elements in coal preparation wastes was performed for the U.S. Environmental Protection Agency and the U.S. Energy Research and Development Administration (now the Department of Energy) (Wewerka et al., 1976b). It was assumed that most of the knowledge concerning trace elements and minerals in raw coal could be logically applied directly to coal wastes. It was also stated that sufficient data exist to show that harmful or toxic quantities of elements (e.g., manganese, cobalt, nickel, and zinc) might be released to the environment in the coal preparation processes.

Reported data and literature suggest that those trace elements that might be a problem in transportation could be largely controlled or removed by the coal washing or preparation process. The major environmental contribution noted has been leachates through or around coal storage piles exposed to weathering.

Rainwater percolating through coal piles may cause local soil contamination depending on the physics and geochemistry of the affected soil. Any water draining from coal storage areas would have to be treated in the same way as drainage water from surface mining operations.

Coal dust seems to be the major pathway for the transport of trace elements blown from open trucks, railroad gondolas, or barges onto surrounding soil or water adjacent to the haul routes. The limited literature on this subject considers it as a physical problem that could be easily controlled by covering the transport carriers, dampening with water, or coating with a chemical film to control removal by the wind or by rain percolation.

Pipeline coal slurry transport was not a concern, as the worst problem encountered might be an accidental spill or leakage that would cause no more than a short-term water-pollution situation (by slurry contamination) rather than a long-term cumulative transport problem for trace elements.

Because there is little literature, data, or information concerning trace elements in transport or storage, it appears likely that these procedures make no significant contribution to the distribution of toxic trace elements in the environment. Coal is a stable, rather safely transported fuel that lacks many of the dangers or problems of leakage, toxicity, explosiveness, flammability, or similar problems encountered with other fuels, such as oil and natural gas. It is doubtful that the storage or transport methods currently in use in the United States will present any severe hazards, because most existing difficulties could be rather easily controlled by feasible and economical methods of wetting, covering, or otherwise limiting potential water-leachate or blowing-dust problems.

HEALTH EFFECTS

Occupational Health

Occupational health impacts from trace metals mobilized during coal transport are unknown, but worker exposure might occur through inhalation of coal dust during loading and unloading of the coal. Estimates of coal dust lost during transport range up to roughly 1 percent of the total tonnage (Council on Environmental Quality, 1973). Losses are from fugitive emissions during loading and unloading and windblown coal dust during transit. Losses vary with type of coal, method of loading and unloading, condition of rail cars, moisture and content of the fines, speed of the train, and wind speed.

These are uncontrolled emission estimates, however. Several methods are available to reduce fugitive dust emissions. These include negative pressure hoods over loading operations, wind guards and covers for gondola cars, and chemically sealing or dust-proofing the surface of each load (Szabo, 1978). However, the extent to which use of such methods reduces total emissions is unknown.

Thermal drying of coal can produce large amounts of dust, and workers at mechanical coal-cleaning plants with thermal drying equipment could be affected. Most thermal drying of coal is done in the Appalachian region, although only 13 percent of coal mechanically cleaned nationwide in 1975 was thermally dried. West Virginia (with 51 percent of the nation's thermal drying plants) accounted for 45 percent of the total, and Virginia (with 11 percent of these plants) accounted for 11 percent (U.S. Bureau of Mines, 1977, p. 438).

Public Health

Transport by Unit Train, Barge, and Truck

Trace metals are emitted during coal transport through diesel-fuel combustion and from windblown coal particles. The large regional differences in volumes of coal transport result in order-of-magnitude or greater differences in emissions of all metals in regions east of the Mississippi River compared with regions west of the Mississippi (see Appendix C). Unit trains transport more tonnage nationally than either barges or trucks (Appendix A), thereby accounting for the largest share of the emissions. In Northern Appalachia, barge and truck emissions combined are greater than unit-train emissions, but unit-train emissions constitute over 50 percent of emissions from coal transport in the Central Appalachia and Central regions. In the Northwest and Southwest regions, trains are the primary mode of coal transport; barge and truck contributions to air pollution are negligible by comparison.

Public health impacts of trace metals mobilized during coal transport are not known. Emissions from low-level sources will be mobilized over shorter distances, with risk of increased exposure for only relatively small populations. Regional emissions range from 10 ton/yr to 10^{-3} ton/yr for various metals (Appendix C).

Table 14 indicates that arsenic may be mobilized to air in roughly equal amounts by the coal combustion process and by transportation. Arsenic in coal is associated with a mineral phase, in pyrite or possibly arsenopyrite (Swaine, 1977), which has physical and chemical properties that are different from the elemental, oxide, and sulfate forms found in combustion products (Bertine and Goldberg, 1971; Davison et al., 1974).

Transport by Slurry Pipelines

Populations living near slurry pipeline routes could be exposed to pollutants mobilized by accidental slurry release, slurry storage containment failure, fugitive dust, or power-use emissions (Mining Information Services, 1977). The probabilities of such exposures occurring and resultant effects have not been quantified and would be confined to local areas.

Coal Cleaning

Trace metals mobilized in air and water from effluents of coal preparation constitute potential public health problems (Wewerka et al., 1976b). The effluents from coal preparation and processing include rock, mineral impurities, and other noncoal solids (known generally as coal refuse) and an aqueous effluent ("blackwater").

Up to 2 tons of blackwater, containing from 4 to 5 percent of suspended coal particles, can be generated per ton of coal washed. Blackwater is usually sent to a tailings pond, where the solids are allowed to settle and the clarified water recirculated. Washing coal with water removes dust and fine organic material adhering to the coal.

TABLE 14 Ratios of Amounts of Trace Elements and Iron Mobilized into Air by Steam-Electric Plant Coal Combustion to Amount Mobilized by Coal Supply Process in 1975, by Metal and Region[a,b]

Metal	Region				Range of Magnitudes
	Appalachia	Central	Northwest	Southwest	
Arsenic	2	1	0.3	0.5	0.1-1
Beryllium	20	40	10	600	10-100
Cadmium	300	3,000	20	7	1-1,000
Chromium	100	300	70	300	10-100
Copper	9	20	9	40	1-10
Iron	10,000,000	40,000,000	300,000	80,000,000	100,000-10,000,000
Lead	100	500	20	700	10-100
Mercury	4,000	2,000	60,000	2,000	1,000-10,000
Nickel	500	400	50	200	10-100
Vanadium	70	60	1,000	200	10-1,000
Zinc	60	60	800	300	10-100

[a]Source: Appendix C.

[b]Includes extraction (underground, strip, and auger) and transportation (unit train, barge, truck). Air emissions from preparation were reported as being zero.

Density separation uses media with different densities to separate the heavier particles. Seventy-five percent of the total commercial production of washed coal used density separation methods, which reduce the concentration of certain trace elements in the washed coal. Aluminum, antimony, arsenic, calcium, cadmium, cobalt, copper, gallium, iron, lead, manganese, mercury, molybdenum, nickel, potassium, selenium, titanium, zinc, and zirconium are susceptible to removal from coal by density separation methods (Wewerka et al., 1976b).

Solid wastes from coal cleaning and processing, an overall average of 23 percent by weight of the mined coal, concentrate the components to constitute the most potent source of AMD in the coal-fuel cycle. All such wastes should be returned to the mining sites for suitable disposal to prevent leaching and other forms of escape into the active environment. The finely divided and well-exposed state of coal refuse enables acid-producing and weathering processes to work more effectively. Thirty percent of AMD from abandoned or neglected mines has been attributed to coal-waste materials. There are 3000 to 5000 abandoned coal-waste piles in the United States. Typically, 1.5 to 2.0 pounds of acid (H_2SO_4) and 0.5 to 0.7 pounds of soluble iron are produced per acre of refuse per day, but rates of more than 300 pounds of acid per acre per day have been reported from highly mineralized areas. As with AMD from mines, water supplies are affected, but the direct impact on public health is not known (NRC Study Committee to Assess the Feasibility of Returning Underground Coal Mine Wastes to the Mined-Out Areas, 1975).

Ignition of Coal-Waste Piles

In the late 1960's, an estimated 300 to 500 coal-waste piles were burning, as a result of intentional ignition or by spontaneous heat generated by reactions within the pile. Air pollution from burning refuse piles is highly concentrated in the vicinity of the sources and can have considerable local impact. Many trace elements are transported from the hot zone by the generated vapors and deposited in cooler surrounding areas. Large amounts of the major pollutants--sulfur oxides, nitrogen oxides, hydrocarbons, carbon monoxide, and particulates--are also emitted from the heaps (Wewerka et al., 1976a).

Burning piles are generally located near small communities. In one survey of 292 burning piles, 45 percent were within less than 1 mile of a community of 200 or more people. Nine percent were in areas populated by 10,000 to more than 100,000 people (McNay, 1971).

The public health impact of trace elements in air pollution from burning refuse heaps is unknown. Whether or not the air-quality standards are exceeded by burning coal-waste piles, the same emission regulations apply as for any uncontrolled burning, regardless of whether started by spontaneous combustion or by other circumstances.

5

COAL COMBUSTION

Estimated total amounts of trace elements mobilized in the environment from coal use in 1975, compared with those mobilized in the form of industrial and consumer products, are presented in Table 15. The data show that the amounts of beryllium, lithium, selenium, and vanadium mobilized through coal use are one to two times greater than those mobilized through industrial and consumer products. The amounts of boron, barium, and cobalt mobilized are 39, 22, and 17 percent, respectively, of those mobilized by industrial and consumer products. Mobilization of antimony, cadmium, fluorine, mercury, and molybdenum from coal use is 3 to 6 percent of the amounts mobilized in the form of industrial and consumer products. Compared with industrial and consumer product mobilization, amounts of chromium, copper, lead, nickel, and zinc mobilized by coal use are quite small, in each instance equivalent to less than 2 percent (Table 15).

COMBUSTION RESIDUES

Residues from coal-fired power-generation plants in general can be assigned to the following categories:

- boiler bottom ash (boiler slag);
- fly ash, captured in emission-control devices and/or emitted to the atmosphere through stacks;
- flue-gas desulfurization (FGD) sludge resulting from the removal of SO_2 from the combustion gases;
- atmospheric gases; and
- wastewaters.

The quantities of wastewater in contact with coal-derived trace elements are small in comparison with those of other waste streams noted previously; however, some of the ash constituents might be dissolved in this water. Figure 12 is a residual-flow diagram showing stages where the various types of residues are generated.

Three categories of boiler are commonly used in power-generation plants. These are stoker-fired boilers, cyclone furnaces, and pulverized-coal-fired (PC-fired) furnaces. The use of stoker-fired

TABLE 15 Estimates of Annual Mobilization of Trace Elements into the Environment from Coal Use and Industrial and Consumer Products

Element	Coal Use[a] (tons/year)	Industrial and Consumer Product Use[b] (tons/year)	Ratio of Coal to Product Use
As	1,669	—	—
Sb	389	12,987	0.030
Ba	166,890	757,000	0.220
Be	389	192	2.03
B	38,941	100,000	0.389
Cd	111	3,340	0.033
Cr	3,894	256,000	0.015
Co	1,112	6,400	0.174
Ca	5,563	1,530,000	0.004
F	35,046	606,000	0.058
Pb	2,782	1,300,000	0.002
Li	3,894	3,540	1.10
Hg	68	1,932	0.035
Mo	834	25,872	0.032
Ni	2,782	146,500	0.019
Se	723	530	1.36
V	8,345	5,500	1.52
Zn	10,570	1,232,000	0.009

[a]For 1975 with coal use of 556,301,000 tons and assuming that concentrations of trace elements in all coal used are equal to mean values for subbituminous coal. From Hittman Associates, Inc. (1977).

[b]1975 consumption. From U.S. Bureau of Mines (1977).

boilers for major electric power generation stations is diminishing; however, their use in smaller industrial plants is expected to increase as these plants convert from oil firing to coal firing. They represent a potential source of widely dispersed, low-level emissions, which are frequently poorly controlled by the operators. High temperatures encountered in the fuel beds could result in volatilization of some elements that are not normally vaporized in large PC-fired furnaces. Construction of new cyclone furnaces in electric power generation stations has essentially stopped because of problems associated with emissions of oxides of nitrogen. In the cyclone unit, 70 to 80 percent of the ash is melted and is continuously tapped from the bottom of the furnace as slag. The remaining 20 to 30 percent is collected from the stack gas by techniques similar to those used in the PC-fired units. The PC-fired boiler unit is of two types--wet and dry bottom--with newer units being almost exclusively the latter. In both, the bottom of the fire box contains one or more hopper-shaped openings through which 20 to

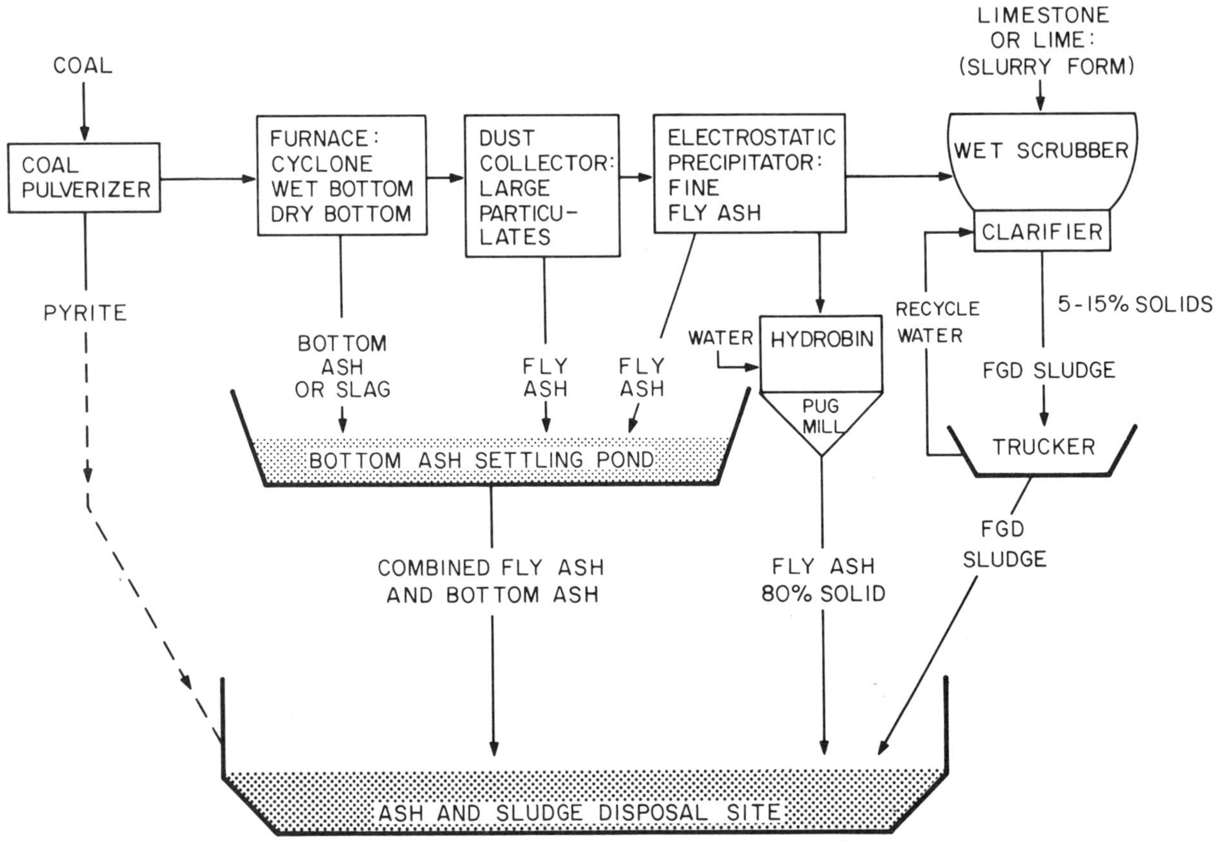

FIGURE 12 Flow diagram of a coal-fired power plant equipped with FGD sludge system (Bern, 1976).

30 percent of the ash in the form of large fused pieces drops into a water-filled hopper, whence they are subsequently sluiced to a settling pond. The wet-bottom furnace, whether PC or cyclone, operates at higher temperatures in the fire box so that more of the ash is melted and flows through a slag tap into a water-filled hopper below. This material is referred to as boiler slag. Small ash particles entrained with the combustion products from the boiler are collected at the stack by means such as electrostatic precipitators and bag filters (fly ash) and wet scrubbers (FGD sludge), with collection efficiencies as high as 95 to 99.7 percent by weight. The percentage of bottom ash, boiler slag, and fly ash in relation to the furnace configuration is shown in Table 16.

The amount of bottom ash, boiler slag, and fly ash produced in the United States in 1971 was estimated at approximately 10, 5, and 27 million tons, respectively (Brackett, 1973). In 1974, ash production soared to a record high of 59.5 million tons (National Ash Association, 1975).

The main uses for these types of coal ashes are manufacturing cement, fill material, and road and airport-runway construction. The amounts of bottom ash, boiler slag, and fly ash used in 1971 were 16.0, 75.2, and 11.7 percent of their respective totals produced that year. Increased coal consumption and increased efficiency of ash collection by the power-generation industries will result in production of greater

TABLE 16 Percentage of Coal Ash from Various Furnace Configurations[a]

Furnace	Bottom Ash (percent)	Boiler Slag (percent)	Fly Ash (percent)
Stoker fired, traveling	100	—	—
Stoker fired, spreader	—	45-85	15-55
Cyclone fired	—	80-85	15-20
Pulverized coal, wet bottom	50	—	50
Pulverized coal, dry bottom	20-25	—	75-80

[a]Source: Bern (1976).

quantities of ash. The need for new utilization and/or proper methods of disposal is evident.

Data on the amount of coal residues emitted to the atmosphere and subsequently deposited on vegetation and soils are rather limited, because of the obvious difficulties involved in such measurements. However, mass balance estimates have been made to measure the approximate amount and/or rate of fine particulate emission. Table 17 gives an example of the percentage of fly ash (ash in the flue gas) in three different power stations. Collection efficiencies better than 99 percent are shown for the noncyclone boiler stations. Estimates made over the past several decades by the U.S. Bureau of Mines suggest that fly ash released to the atmosphere is about 10 percent of the total amount of ash in (the combusted) coal. The estimated amounts of 25 trace elements discharged into the atmosphere from three coal-fired steam plants with a combined daily coal consumption of 25,000 tons are shown in Table 18.

The need to remove substantial quantities of SO_2 from the flue gases is dictated by the Federal Clean Air Act and by state and local regulations. This legislation requires the power industry to use

TABLE 17 Percentage of Fly Ash in the Flue Gas at Three Power Stations[a]

Station Description	Fly Ash (percent of total ash)
1. Wyoming, subbituminous coal, 750 MW	0.3
2. Wyoming, subbituminous coal, 350 MW	0.7
3. North Dakota, lignite coal, 250 MW	12.9[b]

[a]Source: U.S. Environmental Protection Agency (1975).

[b]This is a cyclone boiler station.

TABLE 18 Combined Daily Atmospheric Discharge of Elements from Coal-Fired Steam Plants[a]

Element	Atmospheric Discharge (kg/day)	
Al	510	
As	3.4	
Ba	5.1	
Br	96	(gaseous)
Ca	170	
Cd	0.2	
Co	0.3	
Cr	5.1	
Cs	0.2	
Fe	3400	
Hg	1.7	(gaseous)
K	153	
Mg	850	
Mn	3.3	
Na	68	
Pb	3.3	
Rb	1.2	
S	150,000	(gaseous SO_2)
Sb	3.3	
Se	6.9	(90% gaseous)
Th	0.2	
Ti	68	
U	0.3	
V	6.8	
Zn	34	

[a]Modified from Andren *et al.* (1974). The data are for three coal-fired steam plants with a combined daily coal consumption of approximately 23,000 metric tons.

special equipment to remove SO_2 from the flue gases. The wet-scrubbing method is currently preferred, in which a slurry of lime or limestone is contacted with the flue gas. Typical annual production of sludge, and for comparison, ash by a 1000-MW power plant equipped with lime/limestone flue-gas desulfurization system is shown in Table 19. By 1985, the estimated annual production of FGD sludge will amount to 37 million tons originating from 42,535 MW capacity having FGD systems in operation (Bern, 1976).

A more recent technology to minimize SO_2 emission from coal combustion uses fluidized-bed combustion boilers where crushed coal is burned on a bed of inert ash and crushed limestone. The lime reacts with SO_2 released during the combustion of the coal to form sulfates. The resulting fluidized-bed boiler (FBB) waste is a granular solid material consisting of $CaSO_4$, unreacted CaO, other metal oxides, and fly ash. As this technology is only in its infancy, no satisfactory estimate of the FBB waste can be projected.

Because the present coal use is dominated by electric power generation, which will also maintain the greatest growth in use, the

TABLE 19 Typical Coal-Residue Production in a 1000-MW Power Station Controlled with Lime/Limestone Flue-Gas Desulfurization (FGD) System[a,b]

Residue	Production (tons)	
	Wet	Dry
Coal ash (80 percent solids)	422,000	338,000
Limestone sludge (50 percent solids)	762,800	383,400
Lime sludge (50 percent solids)	622,400	311,200
TOTAL	1,807,200	1,032,600

[a]Source: Bern (1976).

[b]The station consumes 2,556,928 tons of coal annually. The coal contains 3 percent sulfur and 13 percent ash; scrubber-removal efficiency is 85 percent.

following discussion will focus on the combustion residual streams outlined in Figure 12. Furthermore, because western coals are expected to be developed more rapidly than eastern coals, emphasis will be on residues from western coals.

Physical Properties

Fly ash and bottom ash are characterized by low specific gravities and bulk densities (Table 20), which make them good structural fill materials and also allow them to be used to form lightweight aggregates and comparatively lightweight building blocks. The bulk density of FGD sludge will vary with its solids content. Limestone sludge has greater dewatering capacity than lime-derived sludge. At the Mojave Generating Station, limestone was selected as the prime reagent in a series of exhaustive tests with 175-MW scrubber modules by the Southern California Edison Company. The limestone sludge has 65 percent solids by settling, 75 percent solids by centrifugation or free drainage, and 80 percent solids by filtration. This distribution results from large amounts of gypsum ($CaSO_4 \cdot 2H_2O$) in the sludge. Lime sludge generated at the Tennessee Valley Authority's Shawnee Generating Station had 45 percent solids by settling, 55 percent solids by free drainage, 57 percent solids by centrifugation, and 63 percent solids by filtration.

Bottom ash is highly permeable, whereas the permeability of fly ash in relation to bottom ash is low. As a result, bottom ash is used as a drainage blanket on various fill structures.

Particle size of bottom ash and boiler slags ranges from coarse gravel to fine sand. They overlap at the fine end with fly ash. The particle size distribution of fly ash and FGD sludge is shown in Table 21. More than 70 percent of the fly-ash mass is made of particles of <53 μm (micrometers, average particle diameter) or particles with diameters equivalent to those of silt (2 to 50 μm) and clay (<2 μm). Rees and Sidrak (1956) have also shown that fly ash derived from British coal sources is made up of particles principally in the silt and clay size range. Cope (1962) also showed that fly ash contains from 30 to 60

TABLE 20 Some Physical Parameters of Coal Residues[a]

Parameters	Fly Ash	Bottom Ash and Boiler Slag	FGD Sludge
Specific gravity	2.1-2.6	2.5-2.6	2.3-2.5
Bulk density (g/cm^3)	1.12-1.28	1.15-1.49	1.42[b]
Permeability (cm/s)	0.05×10^{-5} to 8×10^{-5}	0.025-0.094	10^{-4}-10^{-8}
Average moisture content (percent)	15	20	82-185
Compaction indices			
Maximum dry density (g/cm^3)	1.19-1.49	1.16-1.87	1.29-1.52
Optimum moisture content (percent)	19.5-32	13.8-26.2	20-29

[a]Source: Bern (1976).

[b]The bulk density of FGD sludge depends on the moisture content of the material. The value given is for a 50 percent moisture sludge.

percent silt-sized particles and from 0 to 4 percent clay-sized particles. The grain size distribution of untreated and combined sludge and fly ash is similar (Table 21). Approximately 95 percent of the particles of this material are of the silt size range, and about 5 percent or less are in the clay size range.

The morphology of coal-ash particles has been examined by a number of investigators (Fisher et al., 1976; Natusch et al., 1975; Paulson and Ramsden, 1970; Straughan et al., 1978). Scanning electron microscopic (SEM) observation of fly ash reported by Fisher et al. (1976) demonstrated that fly ash is composed of large, hollow, spherical, aluminosilicate particles called plerospheres (or cenospheres), which have smaller particles attached to their surfaces and enclosed within their cavities. SEM studies conducted by Page et al. (1979) on fly-ash particles in the size range 1 to 53 μm obtained from the Mojave Generating Station (Figure 13) confirmed the observations made by others on the plerospherical structure of the material. However, plerospheres were absent when particle diameters were less than 8 μm. Particles in this range showed a strong tendency toward agglomeration with smaller particles of submicrometer size attached to their surfaces.

Size and shape of coal-ash particles are important not only for engineering considerations but also from the standpoint of precipitation efficiency in power plants, chemical reactivity, solubility of elements contained in the ash, transportability in the atmosphere, and rates of deposition. Precipitation efficiency in power plants is increased when particles are spherically shaped and when the particle diameter is above 5 μm (Paulson and Ramsden, 1970). Small particles have larger total surface areas available for chemical reactions, travel longer distances, and have longer residence time in the atmosphere. Inclusion of smaller

TABLE 21 Particle Size Distribution of Fly Ash and Flue-Gas Desulfurization (FGD) Sludge

Fly Ash[a]		FGD Sludge[b]	
Particle Diameter (μm)	Mass Fraction (percent)	Particle Diameter (μm)	Mass Fraction (percent)
Wet Sedimentation		Untreated	
>50	32.5	74-2000	2
2-50	63.2	2-74	95
<2	4.3	<2	3
Dry Sieving, Centrifugation and Light Microscopy[c]		1:1 FGD Sludge/Fly Ash	
>1000	0.1	74-2000	1
500-1000	2.6	2-74	94
250-500	1.8	<2	5
105-250	7.2		
53-105	16.2		
30-53	16.5		
12-30	11.5		
2-12	36.1		
<2	6.5		

[a] Source: Page *et al.* (1979).

[b] Source: Bern (1976).

[c] Particles of >53 μm were separated by dry sieving, and those of <53 μm were separated by dry centrifugation and their size estimated by light and electron microscopy.

particles within cavities of larger particles may reduce the solubility of the former when fly-ash particles come in contact with water.

Chemical Composition

The chemical composition of coal combustion ash is influenced by the composition of the source coal and by plant operating conditions. Compositions of different particle sizes of each from a given plant may differ, and the residues are causally related to the ash, sulfur, and calcium content of the coal consumed. As was shown in Table 13, the process of coal cleaning removes a number of trace elements from coal, which alters the chemical composition of the residues.

Influence of Operating Conditions

The vaporization temperatures of the highly volatile elements (e.g., mercury, arsenic, cadmium, selenium, antimony, and thallium) vary according to whether the element is in an elemental or an oxidized state (Table 22). Thus, temperature and oxidation potential of the furnace are important factors in determining trace-element enrichment or depletion in the various types of residues produced. The percentage of

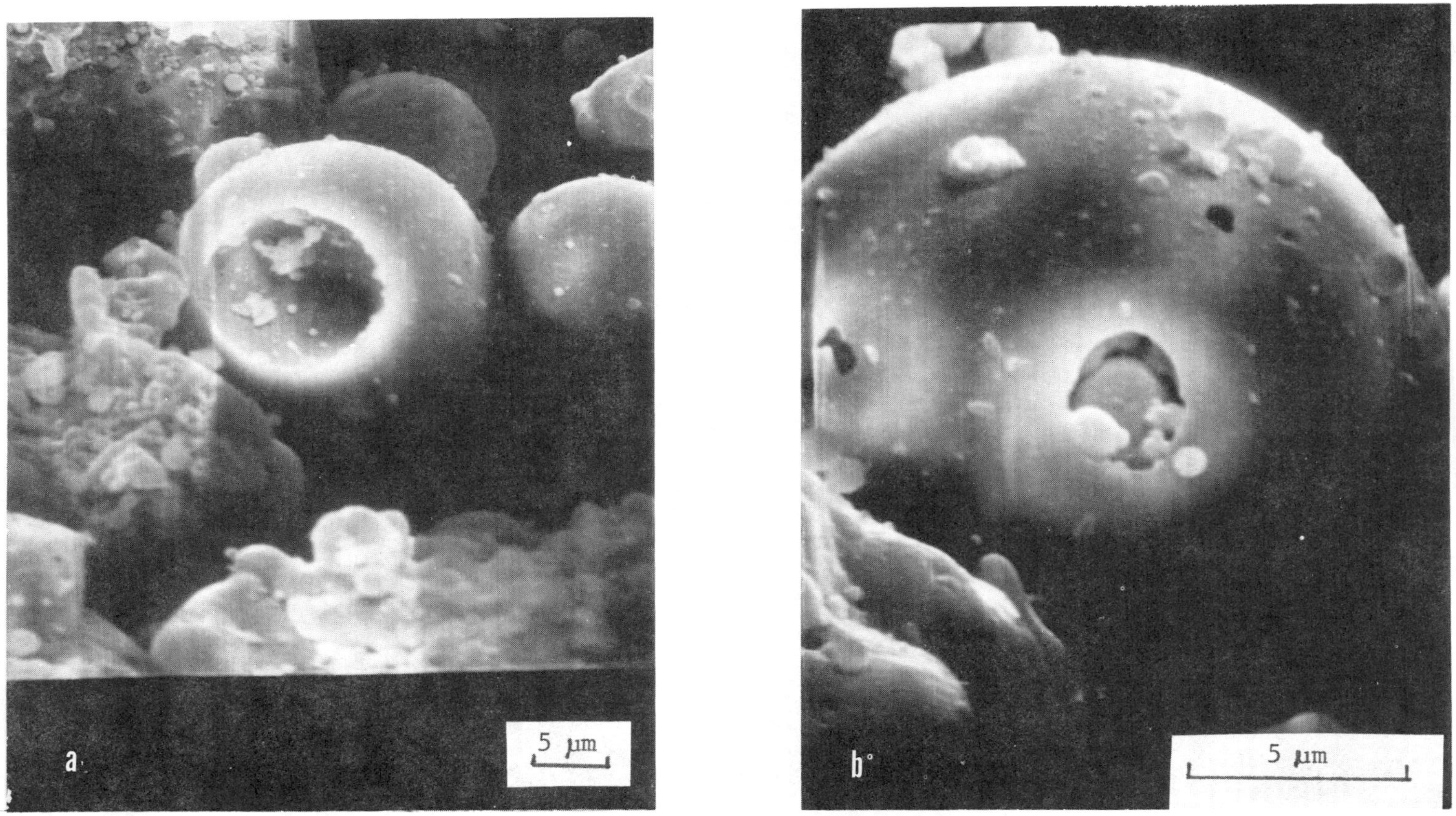

FIGURE 13 Plerosphere with submicrometer spherical inclusions and attached surface spheres: (a) 2000×, (b) 6000× (Page *et al.*, 1979).

elements that are introduced with the coal and retained in each of the three major coal residues is shown in Table 23. An enrichment is indicated when the fraction of the element exceeds that of the residue itself. It is worth noting that about 80, 98, and 88 percent, respectively, of the total mass of chlorine, mercury, and sulfur entering the stream from coal are lost to the atmosphere in the flue gas. Kaakinen et al. (1975) have shown that as residues pass through stages of the power-plant configuration from bottom ash to inlet fly ash to outlet fly ash, progressive increases in the concentrations of copper, zinc, arsenic, molybdenum, antimony, lead, selenium, and mercury occur (Table 24). These authors also showed that concentrations of aluminum, iron, rubidium, strontium, yttrium, and niobium were essentially equal in all combustion residues. Observations by Block and Dams (1976) on ashes derived from Belgian coal sources indicated that magnesium, aluminum, calcium, scandium, titanium, chromium, iron, cobalt, rubidium, barium, and mercury were depleted in the ash in relation to the original coal composition. However, chlorine, zinc, copper, arsenic, selenium, bromine, antimony, iodine, and mercury were enriched in the fly ash. Because of selective volatilization and subsequent elemental enrichment in particles emitted from stacks, Bertine and Goldberg (1971) estimated that the amounts of arsenic, mercury, cadmium, tin, antimony, lead, zinc, thallium, silver, and bismuth mobilized to the atmosphere from fossil-fuel burning are more than 20 times greater than would be predicted from the concentration of the particular element in the coal.

A relation between particle size and the chemical composition of coal residues has been confirmed by a number of investigators. In all investigations, an inverse relationship is observed between the concentrations of certain elements and particle size. Davison et al. (1974) demonstrated that lead, thallium, antimony, cadmium, selenium, arsenic, zinc, nickel, chromium, and sulfur were markedly increased with decreasing particle size of ashes derived from Indiana coal sources fractionated in the range of 0.65 to >74 μm. Similar observations on boron, cadmium, chromium, copper, manganese, nickel, lead, and uranium were made by Lee and von Lehmden (1973) in the size range of 1.5 to 25 μm. The concentrations of gallium, germanium, mercury, and lead showed an approximately threefold increase between the coarse (>50 μm) and fine (<2 μm) fly ash of Australian coal sources (Swaine, 1977).

Fly ash from the Mojave Generating Station was fractionated into particles with diameters in the ranges of >250 and <50 μm (Page et al., 1979). Similar trends were manifested with respect to dependence of the chemical composition of fly ash on particle size. In this study, 29 elements (arsenic, boron, cadmium, cerium, cesium, chromium, cobalt, copper, gallium, hafnium, lanthanum, manganese, molybdenum, nickel, lead, antimony, scandium, rubidium, selenium, thallium, zinc, zirconium, neodymium, samarium, ytterbium, europium, lutetium, terbium, and uranium) were reported in fractions of <53 and >250 μm. The <53-μm fraction was also fractionated into particles with effective diameters in the ranges of 50 to 20, 20 to 13, 13 to 8, 8 to 3, and 3 to 1 μm. Compared with the 53- to 20-μm fraction, the 3- to 1-μm fraction showed substantial enrichment of calcium, sulfur, sodium, phosphorus, zinc, molybdenum, copper, nickel, cadmium, and lead (Page et al., 1979).

TABLE 22 Boiling Points and Relative Order of Volatility of Inorganic Species Possibly Evolved During Coal Combustion[a]

Species Boiling or Subliming Below 1550°C	Species Boiling or Subliming Above 1550°C
As, As_2O_3, As_2S_3	Al, Al_2O_3
Ba	BaO
Bi	BeO
Ca	Bi_2O_3
Cd, CdO, CdS	C
$Cr(CO)_6$, $CrCl_3$, CrS (1550°C)	CaO
K	Co, CoO, CoS
Mg	Cr, Cr_2O_3
$Ni(CO)_4$	Cu, CuO
$PbCl_2$, PbO, PbS	Fe, Fe_2O_3, Fe_3O_4, FeO
Rb	MgO, MgS
Se, SeO_2, SeO_3	Mn, MnO, MnO_2
Sb, $Sb_2S_3Sb_2O_3$	Ni, NiO
SnS	Pb (1620-1750°C)
Sr	Si, SiO_2
Tl, Tl_2O, Tl_2O_3	Sn, SnO_2, SrO
Zn, ZnS	SrO
	Ti, TiO_2, TiO
	U, UO_2
	ZnO

Relative Order of Volatility

Oxides, sulfates, carbonates, silicates, and phosphates

$As \simeq Hg > Cd > Pb \simeq Bi \simeq Tl > Ag \simeq Zn > Cu \simeq Ga > Sn > Li \simeq Na \simeq K \simeq Rb \simeq Cs$

Elemental state

$Hg > As > Cd > Zn > Sb \geq Bi > Tl > Mn > Ag \simeq Sn \simeq Cu > Ga \simeq Ge$

Sulfides

$As \simeq Hg > Sn \simeq Ge \geq Cd > Sb \simeq Pb \geq Bi > Zn \simeq Tl > Cu > Fe \simeq Co \simeq Ni \simeq Mn \simeq Ag$

[a] Sources: Modified from Davison *et al.* (1974) and Bertine and Goldberg (1971).

Table 25 illustrates the effect of particle size of fly ash on the concentration of 17 trace elements as reported by Ondov et al. (1976). The range of concentration increase in the <3-μm fraction as compared with the >15-μm fraction was from 1.5-fold for manganese to 11.5-fold for cadmium. In this study (Ondov et al., 1976), the authors found little or no enrichment for aluminum, calcium, iron, potassium, sodium, nickel, titanium, magnesium, cerium, cesium, dysprosium, europium, hafnium, lanthanum, neodymium, rubidium, scandium, samarium, strontium, tantalum, terbium, thallium, and ytterbium in this size range.

The availability of elements in coal fly ash is not only a function of particle size and concentration but also of depth in individual coal fly-ash plerospheres. Linton et al. (1976) have shown that beryllium, carbon, calcium, chromium, potassium, lithium, manganese, sodium, phosphorus, lead, sulfur, thallium, vanadium, and zinc were

TABLE 23 Percentage of Elements Entering with Coal Discharged in Various Coal Residues[a,b]

Element	Sluice Ash[c] (22.2%)	Precipitator Ash[d] (77.1%)	Flue Gas[e] (0.7%)
Aluminum (Al)	20.5	78.8	0.7
Antimony (Sb)	2.7	93.4	3.9
Arsenic (As)	0.8	99.1	0.05
Barium (Ba)	16.0	83.9	<0.09
Beryllium (Be)	16.9	81.0	<2.0
Boron (B)	12.1	83.2	4.7
Cadmium (Cd)	<15.7	80.5	<3.8
Calcium (Ca)	18.5	80.7	0.8
Chlorine (Cl)	16.0	3.8	80.2
Chromium (Cr)	13.9	73.7	12.4
Cobalt (Co)	15.6	82.9	1.5
Copper (Cu)	12.7	86.5	0.8
Fluorine (F)	1.1	91.3	7.6
Iron (Fe)	27.9	71.3	0.8
Lead (Pb)	10.3	82.2	7.5
Magnesium (Mg)	17.2	82.0	0.8
Manganese (Mn)	17.3	81.5	1.2
Mercury (Hg)	2.1	0	97.9
Molybdenum (Mo)	12.8	77.8	9.4
Nickel (Ni)	13.6	68.2	18.2
Selenium (Se)	1.4	60.9	27.7
Silver (Ag)	3.2	95.5	1.3
Sulfur (S)	3.4	8.8	87.8
Titanium (Ti)	21.1	78.3	0.6
Uranium (U)	18.0	80.5	1.5
Vanadium (V)	15.3	82.3	2.4
Zinc (Zn)	29.4	68.0	2.6

[a]Sources: Schwitzgebel *et al.* (1975) and U.S. Environmental Protection Agency (1975).

[b]At a 350-MW station equipped with an electrostatic precipitator burning subbituminous coal at the rate of 124 metric tons/h.

[c]Bottom ash sluiced to a settling pond wire.

[d]Ash obtained by electrostatic precipitator.

[e]Fly ash in the flue gas.

preferentially concentrated on the particle surface as compared with their concentrations at a distance inside the particle 500 Å from the surface; however, the nature of the surface of particles is not known with certainty. Collin (1974) cited by Swaine (1977) suggested that on each particle there are three more or less distinct outer layers. The innermost of these layers mainly consists of calcium compounds, the outermost layer is acidic, consisting mainly of sulfuric acid. Between these layers there is a neutral region, intermediate in composition between the outermost and the innermost layer.

Davison et al. (1974) suggested that under very high temperatures of combustion (~1600°C), volatilization of trace elements occurs followed by preferential condensation or adsorption onto the smallest particles.

TABLE 24 Progressive Trace-Element Enrichment in a Coal-Fired Power Plant (ppm)[a]

Sample	Cu	Zn	As	Mo	Sb	Pb	Se	Hg
Coal	9.6	7.3	—	0.99	—	—	1.9	0.070
Bottom ash	82	58	15	3.50	2.8	<5	7.7	0.140
Precipitator ash (inlet)	230	250	120	41.00	14.0	66	27	0.310
Precipitator ash (outlet)	320	370	150	60.00	18.0	130	62	—

[a]Source: Kaakinen *et al.* (1975). Samples were collected at a 180-MW unit, Public Service Company, Boulder, Colorado.

Thus, the magnitude of the surface area of particles affects the amounts of the trace element retained by such mechanism. Swaine (1977) estimated that a typical value for the surface area of fly ash is of the order of 1 m^2/g. Surface areas of bottom ash, inlet precipitator ash (ash entering the precipitator in the flue-gas stream), and outlet

TABLE 25 Effect of Fly-Ash Particle Size on the Concentration of Some Trace Elements (ppm)[a]

	Size Range (μm)			
Element	>15	8-15	3-8	<3
As	13.7	56	87	132
Be	6.3	8.5	9.5	10.3
Cd	0.4	1.6	2.8	4.6
Co	8.9	16.3	19	21
Cr	28	49	59	63
Cu	56	89	107	137
Ga	43	116	140	178
Mn	207	231	261	317
Mo	9.1	28	40	50
Ni	25	37	44	40
Pb	73	169	226	278
Sb	2.6	8.3	13	20.6
Se	19	59	78	198
U	8.8	16	22	29
V	86	178	244	327
W	3.4	8.6	16	24
Zn	71	194	304	550

[a]Source: Ondov *et al.* (1976).

precipitator ash (ash exiting the precipitator in the flue-gas stream) measured by Kaakinen et al. (1975) were, respectively, 0.38, 3.06, and 4.76 m^2/g. These authors indicated that enrichment of trace elements in coal combustion is determined by the physicochemical properties of the elements and their chemical compounds in coal and combustion products, the nature of the coal-burning process, and chemical reactions occurring in the emission-control devices. An enrichment ratio (R_{ij}) of each element (i) in each outlet stream (j) relative to the total outlet can be defined by the equation:

$$R_{ij} = \frac{M_{ij} / A_j}{\sum_{j=1}^{6} M_{ij} / \sum_{j=1}^{6} A_j}$$

where M_{ij} is the flow rate of trace element i in outlet stream j; A_j is the flow rate of aluminum in that stream, and n is the total number of outlet streams. The value of the enrichment ratio indicates whether an element is enriched ($R_{ij} > 1.0$), nonenriched ($R_{ij} = 1.0$), or depleted ($R_{ij} < 1.0$) in an outlet stream relative to aluminum and relative to the total condensed outlet. Aluminum is chosen as a reference element because it is nonvolatile at furnace temperatures and its concentrations in inlet and outlet fly ashes are approximately equal. Enrichment ratios were compared to the various measures of element volatility, melting points, boiling points, and vapor pressures of elemental and oxide forms. Enrichment ratios based on the oxide forms showed excellent agreement with volatility of these forms.

Solid and Slurry Residues

The inorganic constituents in coal and associated mineral matter are redistributed among the bottom ash, fly ash, combustion gases, and the FGD sludge, as a result of coal combustion. The primary product of breakdown of clay is mullite ($Al_6Si_2O_{13}$), of pyrite it is ferric oxide, and of calcite it is calcium oxides, while quartz and some silicates remain unchanged. Quartz, mullite, gypsum, and iron oxides have been identified in fly ash (Page et al., 1979; Swaine, 1977).

The chemical composition of bottom ash and FGD sludge produced in a western power station in relation to the chemical composition of the source coal is shown in Table 26. Concentrations of antimony, bromine, cadmium, chlorine, erbium, copper, fluorine, iodine, selenium, silver, tin, and tungsten in bottom ash were either similar to, or less than, those found in the source coal. Manganese was somewhat unique, showing a concentration in bottom ash 41 times greater than in coal. Concentrations of zinc, lithium, and hafnium were from 11.5 to 18.4 times more concentrated in bottom ash than coal; those of strontium, beryllium, boron, cesium, gadolinium, samarium, scandium, terbium, thorium, thulium, and uranium from 5 to 10 times; and the remaining elements from 1 to 5 times more concentrated in bottom ash than in coal. In general, the concentrations of most elements in the solids of FGD sludges did not differ by more than a factor of 2 (more or less) from

TABLE 26 Chemical Composition of Bottom Ash and FGD Sludge (ppm) and Element Enrichment Ratios in a 330-MW Subbituminous Coal-Fired Power Plant[a]

			Enrichment Ratio[b]		
Element	Bottom Ash	FGD Sludge (Solids)	Bottom Ash / Coal	FGD Sludge / Coal	FGD Sludge / Bottom Ash
Mn	3300	2200	41	28	0.7
P	1800	1700	2.8	2.7	0.9
Sr	970	860	7	6	0.9
Sb	0.49	0.82	1.1	1.9	1.7
As	5.4	17	4.2	13.1	3.2
Ba	320	110	2.5	0.9	0.3
Be	11	7.5	5.9	3.9	0.7
B	800	530	7.3	4.8	0.7
Br	1.5	0.98	0.9	0.6	0.7
Cd	0.67	0.89	1.1	1.4	1.3
Ce	200	100	3.6	1.8	0.5
Cs	4.3	2.8	6.5	4.2	0.7
Cl	26	17	0.3	0.2	0.7
Cr	110	180	3.8	6.2	1.6
Co	26	17	4.4	2.9	0.7
Cu	120	340	4.8	13.6	2.8
Dy	4.2	2.8	2.6	1.8	0.7
Er	0.48	0.48	0.9	0.9	1.0
Eu	1	0.69	1.2	0.8	0.7
F	91	300	0.3	0.9	3.3
Gd	0.69	1.1	5.3	8.5	1.6
Ga	27	48	7.3	12.9	1.8
Ge	2.8	4.0	4.9	7.0	1.4
Hf	5.7	1.8	18.4	5.8	0.3
Ho	0.31	0.20	2.1	1.3	0.6
I	0.6	0.2	0.5	0.2	0.3
La	31	20	3.4	2.2	0.7
Pb	8	27	2.8	9.3	3.4
Li	860	2	12.8	0.03	0.002
Lu	0.15	0.10	1.9	1.3	0.7
Mo	35	23	2.4	1.6	0.9
Nd	130	84	1.7	1.1	0.7
Ni	28	19	2.2	1.5	0.7
Nb	18	24	3.3	4.4	1.3

TABLE 26 (continued)

Element	Bottom Ash	FGD Sludge (Solids)	Enrichment Ratio[b] Bottom Ash / Coal	FGD Sludge / Coal	FGD Sludge / Bottom Ash
Pr	11	7.6	1.6	1.1	0.7
Rb	23	32	2.6	3.6	1.4
Sm	1.9	1.3	5.8	3.9	0.7
Sc	80	53	6.2	4.1	0.7
Se	0.87	5.8	0.5	3.6	6.7
Ag	0.14	0.19	0.6	0.8	1.4
Te	0.16	0.09	1.6	0.9	0.6
Tb	0.46	0.15	5.1	1.7	0.3
Th	13	18	5.0	6.9	1.4
Tm	0.26	0.12	6.5	3.0	0.5
Sn	2.0	3.1	1.1	1.7	1.6
W	3.7	5.0	1.1	1.5	1.4
U	23	7.2	9.2	2.9	0.3
V	180	280	2.5	3.8	1.6
Yb	1	1.4	3.7	5.2	1.4
Y	75	50	4.4	2.9	0.7
Zn	10	66	11.5	76	6.6
Zr	62	76	2	2.5	1.2

[a]Source: U.S. Environmental Protection Agency (1975).

[b]Ratios are calculated on the basis of the chemical composition of coal used at this station.

those of the bottom ash. Exceptions were arsenic, bismuth, copper, fluorine, lead, selenium, and zinc, which were concentrated by more than a factor of 2 in FGD sludge compared with bottom ash, and barium, hafnium, iodine, lithium, terbium, and uranium, which were concentrated by more than a factor of 2 in bottom ash than in FGD sludge.

Table 27 gives the average and range of the elemental composition of fly ash and soils. The major components of fly ash--silicon, aluminum, calcium, magnesium, iron, sodium, potassium, and titanium--make up from 35 to 70 percent of the total elemental composition of fly ash. Fisher et al. (1976) indicated that the composition of fly ash from western U.S. coals could be approximated by the formula,

$$Si_{1.00}Al_{0.45}Ca_{0.051}Na_{0.047}Fe_{0.39}Mg_{0.020}K_{0.017}Ti_{0.011}.$$

TABLE 27 Chemical Composition of Fly Ash and Soil[a]

Element	Range		Average	
	Fly Ash	Soil	Fly Ash	Soil
Element (percent)				
Al	0.1-17.3	4-30	11.7	7.1
Si	19.1-28.6	25-33	26.6	33
Ca	0.11-12.6	0.7-50	4.4	1.4
Mg	0.04-6.02	0.06-0.6	0.9	0.5
Na	0.01-0.66	0.04-3.0	0.6	1.4
K	0.19-3.0	0.04-3.0	0.8	1.4
Fe	1-26	0.7-55	4.1	3.8
Mn	0.01-0.3	0.01-0.4	0.02	0.09
S	0.1-1.5	0.01-0.2	0.2	0.08
P	0.04-0.8	—	0.13	0.07
Sr	0.03-0.3	0.05-0.4	0.09	0.03
Ba	0.011-0.5	0.01-0.3	0.3	0.05
Ti	0.16-0.7	—	0.5	0.5
Element (ppm)				
Sb	1.6-202	0.6-10	1.6	6
As	2.8-6300	0.1-40	9	6
Be	3-7	0.1-100	5.5	6
B	48-618	2-100	390	10
Br	0.7-5.3	—	1.8	25
Cd	0.7-130	0.01-7	1.0	0.06
Ce	22-133	—	150	50
Cs	—	—	5	—
Cl	—	10-1000	47	—
Cr	10-690	5-300	54	100
Co	7-49	1-40	12.6	8
Cu	14-1000	2-100	63	26
F	100-610	30-300	201	200
Gd	—	—	1.4	1.0
Ga	15-93	15-70	33	30
Ge	—	—	3.3	1.0
La	12-72	—	70	30
Pb	7-279	2-100	48	10
Li	50-1064	1-2000	70	30
Mo	7-117	0.2-5.0	10	2
Ni	10-4300	10-1000	34	40
Nb	—	—	20	21
Rb	—	30-600	43	100
Sc	3.7-141	10-25	15.6	7
Se	0.2-134	0.1-2.0	7.6	0.2
Ag	0.04-5	0.1-8.0	0.3	0.1
Te	0.02-0.2	—	0.2	10.1
Tl	0.8-1.7	—	1.2	5.0
Sn	3.3-63	0.1-300	3.3	10.0
V	50-1000	20-250	119	100
Zn	36-1333	10-300	72	50
Zr	50-1286	60-2000	200	300

[a]The range of fly-ash composition represents literature values taken mainly from Block and Dams (1975), Bern (1976), Swaine (1977), Martens (1971), Mulford and Martens (1971), Martens *et al.* (1970), Doran and Martens (1972), and Page *et al.* (1979). The average composition of fly ash was taken from data of Swanson (1972), Schwitzgebel *et al.* (1975), and Page *et al.* (1979) for ashes derived from western U.S. coal sources. Soil composition was derived mainly from Bowen (1966) and Lisk (1972).

Based on the average chemical composition of soils shown in Table 27, application of fly ash to agricultural soils will probably result in soil enrichments of calcium, barium, strontium, sulfur, boron, molybdenum, selenium, and lead. Limited enrichments of other elements are also possible, depending on the rate of application, climate, and type of soil and fly ash. Of the trace elements shown, boron, molybdenum, and selenium are of concern. Boron is toxic to plants in relatively small concentrations in soil, and molybdenum and selenium are toxic to animals when present above some critical concentrations in the plant tissue.

Water Extracts from Residues

As the basic concern is the availability of trace elements to the biological systems, measurement of water solubility of the elements is used to reflect this bioavailability. The major- and trace-element contents in saturation extracts (extracts obtained from soils saturated with water) from fly ash obtained at the Mojave Generating Station compared with those for 68 soils from California as reported by Bradford et al. (1971) are shown in Table 28. The high pH of saturation extracts from fly ash was responsible in part for the relatively low solubility of most elements. Concentrations of calcium, sodium, barium, strontium, and molybdenum were higher in the fly-ash extracts than in the soil extracts. Barium and strontium contents of the pH 12.5 fly-ash extract were more than 50-fold greater than the mean values of these two elements in 68 soil extracts, suggesting that saturation extract concentrations of barium and strontium, in certain source- and site-specific instances, may serve as a good indicator of the extent of fly-ash contamination of soils (Bradford et al., 1978).

Acidifying the ash to pH 6.5 greatly increased the solubility of most elements (Table 28). Concentrations of calcium, magnesium, strontium, boron, manganese, and molybdenum in the acidified fly-ash extract were considerably greater than their corresponding concentrations in saturation extracts of fly ash without acidification. The pH of soils, in general, ranges between 4.0 and 8.2. Because soils usually have a high capacity to resist changes in pH, if fly ash were mixed with soil in amounts equivalent to a few percent by weight, the resultant equilibrium pH should fall in the range common to soils.

Table 29 shows the composition of leachates from coal ash and FGD sludge of a western power station. For comparative purposes, water-quality criteria for public supplies and irrigation water are also indicated in Table 29. Leachates from a mixture of fly ash and bottom ash (4:1) contain excessive concentrations of barium, molybdenum, and selenium, compared to standards for public water supplies and/or irrigation water. Concentrations of boron, fluorine, molybdenum, and selenium in the FGD-sludge leachates exceed those established for public water supplies and/or irrigation waters. The data presented in Table 29 indicate that those elements present in residues from coal combustion that are most apt to contaminate water supplies are barium, boron, fluorine, molybdenum, and selenium.

TABLE 28 A Comparison of Major and Trace-Element Contents in Saturation Extracts of Mojave Fly Ash and 68 California Soils[a]

Element	Fly Ash (μg/ml) Water Soluble Before pH Adjustment (pH 12.5)	Fly Ash (μg/ml) Water Soluble After pH Adjustment (pH 6.5)	Concentration in Soil Saturation Extracts (μg/ml) Mean	Concentration in Soil Saturation Extracts (μg/ml) Median
Ca	476	38,234	128	60.0
Mg	<1	849	38	12.4
Na	287	900	524	45.0
K	<100	<100	20	10.0
Si	<0.6	<0.6	3.1	5.0
B	<0.6	65	3.1	<0.1
Ba	50	15	0.26	0.10
Sr	61	333	0.93	0.18
Al	<2	<2	0.40	<0.01
Cr	<1	<1	0.01	<0.01
Fe	0.01	0.11	0.05	0.03
Mn	<0.01	1.3	0.17	<0.01
Cu	0.01	0.02	0.04	0.03
Zn	0.02	0.08	0.07	0.04
Mo	0.12	1.11	0.73	<0.01
Ni	0.01	0.13	0.02	<0.01
Co	<0.01	0.09	0.06	<0.01
V	<0.01	0.02	0.07	0.01
Pb	<0.02	<0.02	0.05	<0.01
Cd	<0.01	0.03	—	—
Ag	<0.001	<0.001	—	—

[a]Source: Bradford *et al.* (1971, 1978).

Atmospheric Deposition

The quantity of trace elements emitted to the atmosphere from stationary sources depends on such factors as concentration in coal, boiler configuration and load, properties of the element and its compounds, and the effectiveness of particle control devices. Paulson and Ramsden (1970) have also indicated that the microlithotype of the source coal plays an important role in determining the precipitation efficiency of the electrostatic precipitators. They pointed out that some coals are rich in fusite (a coal microlithotype), which results in production of fine fly ashes that are predominantly of submicrometer size, hence difficult to precipitate. With the use of a cascade impacter, Lee et al. (1975) were able to show that between the inlet and the outlet of an electrostatic precipitator, the fraction of particles of >5 μm decreased

TABLE 29 Comparison of Trace-Element Concentrations of Leachates from Coal Ash (a Mixture of 80 Percent Fly Ash and 20 Percent Bottom Ash) and FGD Sludge to the Recommended Maximum Concentrations for Public Water Supplies and Irrigation Waters (mg/liter)

Element	Concentration in Leachate from[a]		Recommended Maximum Concentration for[b]	
	Coal Ash	FGD Sludge	Public Water Supplies	Irrigation Water[c]
Arsenic	<0.002	<0.002	0.10	0.10
Barium	40	2.0	1.0	—
Beryllium	0.003	0.002	—	0.10
Boron	0.03	2.6	—	0.75
Cadmium	<0.001	0.0005	0.01	0.01
Chromium	<0.001	0.001	0.05	0.10
Fluorine[d]	2.3	31.5	1.4-2.4	1.0
Mercury	0.0006	0.0005	0.002	—
Lead	0.0068	0.0056	0.05	5
Manganese	<0.002	<0.002	0.05	0.20
Molybdenum	0.047	0.063	—	0.01
Nickel	<0.05	<0.05	—	0.20
Selenium	0.009	0.045	0.01	—
Vanadium	<0.10	<0.1	—	0.10
Zinc	0.038	0.005	5.0	2.0

[a]Source: Radian Corporation (1975). Data are for a Southwest power station burning low-sulfur coal.

[b]Source: U.S. Environmental Protection Agency (1973).

[c]Recommended maximum concentrations of trace elements in irrigation waters used for sensitive crops on soils with low capacities to retain the elements in unavailable forms.

[d]Recommended maximum concentration depends on mean annual temperature of the region; as temperature increases maximum concentrations decrease.

from 96 to 48 percent, while that of the <5-μm particles increased from 4 to 52 percent. They also showed that chromium, lead, antimony, zinc, and selenium were predominantly in the 0- to 1-μm size range; cadmium, iron, and vanadium in the 1- to 5-μm range; nickel in the 5- to 10-μm range, and manganese was completely removed by the control device. Zoller et al. (1975) estimated that the amount of fly ash emitted from the stack of two 710-MW coal-burning units consuming 116 tons/h of coal with 10 percent ash and 1 percent sulfur was on the order of 1.2 tons/h. This amounts to about 1 percent of the amount of coal used or about 10 percent of the amount of ash produced.

Klein and Russell (1973) found that soils around a coal-burning facility (650 MW, 90 percent precipitator efficiency) were enriched in silver, cadmium, cobalt, chromium, copper, iron, mercury, nickel, titanium, and zinc. They also found that plant materials (native grass, maple leaves, and pine needles) were enriched in cadmium, iron, nickel, and zinc. Soil enrichment, except for mercury, was correlated with wind patterns and metal content of coal. Bradford et al. (1971), however, concluded that four years of operation at the Mojave Generating Station resulted in no measurable contamination of either soil or vegetation in the region surrounding the facility.

Anderson and Smith (1977) examined concentrations of mercury in the surface 2 cm of soil within 20 km of a 1200-MW coal-burning facility (Table 30). Soils sampled in the northerly directions (prevailing downwind) contained higher concentrations than those sampled in the southerly directions. Despite some increased concentrations of mercury, however, the projected soil enrichments of the element over a 35-year period are not critical.

A mass deposition model was formulated by Jurinak et al. (1977) for particulates emitted from a hypothetical 3000-MW coal-burning facility situated in a semiarid environment. Some of the basic assumptions of this model are emission rate, 1.05 tons/h; deposition velocity, 0.03 to 0.1 m/sec; particle size distribution for particles >10 μm that was 3 percent of 25-μm radius, 5 percent of 12.5-μm radius, and 10 percent of 7.5-μm radius. Deposition rates established by this model are shown in Figure 14 for the winter, spring, summer, and fall seasons. The changes in pattern of deposition with season were modeled to reflect changes in the wind speed and direction and atmospheric stabilities. Deposition was found to be highest in summer and lowest in fall. Total particulate mass deposition ranged from 10 to 50 kg km^{-2} $month^{-1}$. The model also predicted that fallout over a 50-year period would be insignificant with

TABLE 30 Mercury Concentrations of Surface 2 cm of Soil in a Region Within 20 km of a Coal-Fired Power Plant in Illinois[a]

	Number of Observations	Mean Mercury Concentration After 7 Years of Operation (μg/g)
Direction from power plant		
NW	15	0.019
SW	20	0.015
NE	21	0.022
SE	17	0.016
Maximum concentration observed		0.037
Projected lifetime enrichment		
Uncultivated soil		0.110
Cultivated soil		0.015

[a]Source: Anderson and Smith (1977), 1200-MW facility. Coal used had a mean Hg concentration of 0.2 μg of Hg/g.

regard to the total amount of zinc, chromium, lead, and cadmium originating from the stack and deposited in the final environmental sink for the region. Fallout, however, would contribute 65 percent of the mercury loading of this sink.

Based on the maximum deposition rate (50 kg km^{-2} $month^{-1}$) found by Jurinak et al. (1977) and concentration of trace elements in the <3-μm fraction (Table 25), the concentration of fly-ash-derived trace elements in plants and soils was estimated by Page et al. (1979). These are shown in Tables 31 and 32 and are compared with their typical values in plants and soils. These estimates show measurable, but small, enrichments in plants or soils in trace elements originating from stack emissions. Lyon (1977) and Vaughan et al. (1975) arrived at similar conclusions using different approaches to estimated fallout.

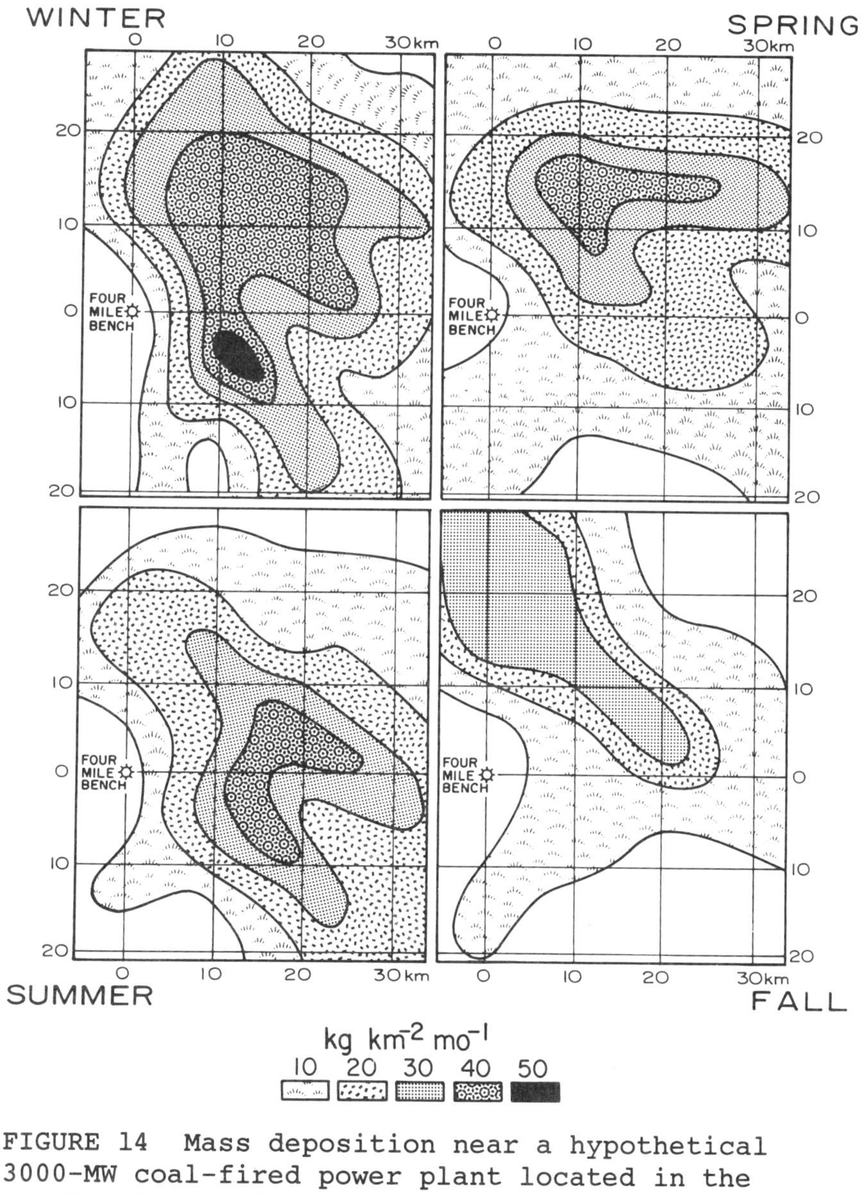

FIGURE 14 Mass deposition near a hypothetical 3000-MW coal-fired power plant located in the southwestern United States (Jurinak et al., 1977).

TABLE 31 Estimated Maximum Deposition of Trace Elements onto Soil Adjacent to a Coal-Fired Power Plant[a] (microgram of element per gram of soil)

Element	Amount Deposited		Common Soil Concentrations	
	Annual	Lifetime[b]	Typical	Range
As	0.00039	0.014	6	0.1-40
Cd	0.000014	0.0005	0.06	0.01-7
Pb	0.00082	0.029	15	2-200
Mo	0.00015	0.0052	2	0.2-5
Se	0.00059	0.021	0.2	0.01-2
U	0.000087	0.003	1	0.1-10
Zn	0.00163	0.057	50	10-300
Sb	0.000061	0.0021	6	2-10
Be	0.00003	0.0011	6	0.1-40
Cr	0.00019	0.0066	100	5-3000
Co	0.000062	0.0022	8	1-40
Cu	0.00041	0.0144	20	2-100
Ga	0.00053	0.0182	30	0.4-300
Ni	0.00012	0.0042	40	10-1000
Th	0.0009	0.0032	5	0.1-12
V	0.00097	0.034	100	20-500

[a]From Page *et al.* (1977) and Ondov *et al.* (1976). 3000-MW power plant, western U.S. coal, electrostatic precipitator efficiency of 99 percent.

[b]Assuming a 35-year lifetime.

Disposal and Recycling of Collected Residues on Land

The collected residues from coal-fired electric power plants have been in the past, and in all probability will continue to be, disposed or recycled on land. Land-disposal techniques utilized include landfill, lagooning, and land spreading. In situations where the residues are spread onto land that supports either agricultural crops or native vegetation, trace elements in fly ash could adversely affect the productivity of the land or the quality of the vegetation grown. Therefore, a comprehensive understanding of the chemical, physical, and biological changes that occur when residues from coal-fired power plants are mixed with soils is essential to evaluate the effect that these changes will have on the productivity of the land and trace-element composition of the vegetation.

TABLE 32 Estimated Trace-Element Contamination of Vegetation Arising from Emission from a Coal-Fired Power Plant[a] (microgram of element per gram of dry matter)

Element	Concentration Deposited on Vegetation[b]	Typical Concentration for Vegetation
As	0.07	0.4
Cd	0.002	0.2
Cr	0.034	1.5
Cu	0.074	10
Pb	0.15	3
Mo	0.027	1
Se	0.107	0.2
Zn	0.30	25
Sb	0.011	0.06
Be	0.006	0.03
Co	0.011	<1.0
Ga	0.096	1.2
Ni	0.021	5
Th	0.016	0.05
U	0.016	0.04
V	0.18	1

[a]Sources: From Page *et al.* (1979), Jurinak *et al.* (1977), and Ondov *et al.* (1976). 3000-MW power plant, western U.S. coal, electrostatic precipitator efficiency of 99 percent.

[b]Assumptions: 100 percent canopy, yield 3770 kg of dry matter per hectare, 4-month exposure, particulates deposited are <3 μm, and all particulates deposited remain with harvested crop.

Chemical Properties of Fly-Ash-Amended Soils

Table 33 presents the pH, electrical conductivity, and soluble calcium plus magnesium, boron, and SO_4-S (sulfur present as SO_4 but expressed as elemental sulfur) in soils amended with Mojave fly ash, at rates ranging up to 8 percent by weight (~180 metric tons/ha) and cropped to brittlebush (Encelia farinosa) under greenhouse conditions for 8 months. The results of a field study conducted by Plank et al. (1975) with two acid soils amended with fly ash at cumulative rates of 144 metric tons/ha, and cropped to corn for 2 years, are given in Table 34.

Application of fly ash, regardless of its source, increased the pH of soils, irrespective of their type. The calcareous soil (Xerollic Calciorthid, fine, loamy, mixed thermic), Arizo (Typic Torriorthient, sandy, skeletal, mixed, thermic), however, buffered this effect over an 8-month period of cropping (Table 33). Acid soils [Redding (Abruptic Durexeralof, coarse, loamy, kaolinitic, thermic), Groseclose (Typic, Hapludult, clayey, mixed, mesic), and Woodstown (Aquic Hapludult, fine,

TABLE 33 Select Chemical Properties of Soils Amended with Fly Ash and Cropped Under Greenhouse Conditions to Brittlebush (*Encelia farinosa)* for 8 months

Fly Ash Added[a] (percent)	EC_e[b] (mmhos/cm)	pH_{sp}[c] Initial	pH_{sp}[c] Final	Soluble[d] Ca + Mg (meq/liter)	Soluble[d] B (μg/ml)	Extractable[e] SO_4 - S (μg/g)
Arizo Calcareous Soil						
0	1.3	8.0	7.9	10.8	2.3	1.3
1	1.2	8.9	7.9	10.0	5.5	55
2	1.5	9.9	8.0	11.1	6.3	129
4	2.2	10.5	7.9	19.3	7.0	223
8	3.2	10.6	7.9	32.0	11.1	388
Redding Acid Soil						
0	1.5	5.5	4.8	7.6	0.4	14
1	2.6	6.9	6.1	19.7	2.8	20
2	2.2	7.7	7.2	21.4	3.7	49
4	3.2	8.7	7.6	30.4	5.2	128
8	3.7	9.9	7.8	46.7	8.0	276

[a] Fly ash obtained from the Mojave Generating Station.

[b] EC_e is the electrical conductivity of the soil saturation extract.

[c] pH_{sp} is the pH of the soil saturation paste. The initial pH is obtained approximately 2 weeks after preparing the soil/fly-ash mixtures, and the final pH is obtained after soils were cropped for 8 months.

[d] Amounts dissolved in saturation extract from soil.

[e] Amounts extracted with 1 *N* ammonium acetate (NH_4OAc).

loamy, siliceous, mesic)] continued to maintain pH levels greater than those of controls over the 8-month growth period (Tables 32 and 33). Analyses of acid soils showed evidence of calcium carbonate formation at rates of fly ash of 2 percent or greater (Page et al., 1979). This finding is probably the result of the high concentrations of hydroxyl ions in fly ash reacting with carbon dioxide released by the roots. The efficiency of fly ash as liming material was most recently compared with that of calcium carbonates by Phung et al. (1978). Results showed that the fly ash used (see Table 28 for chemical composition) had a neutralizing capacity 35 to 50 percent that of commercial limestone. The authors showed that fly ash and limestone are effective in increasing the levels of soluble and exchangeable calcium and in reducing the concentrations of exchangeable aluminum and manganese to levels not toxic to plants. The soil used in the study was a Reyes silt clay (Sulfic Haplaquept, fine, mixed, acid, thermic) that required the application of an amount of fly ash equivalent to 115 metric tons/ha to raise its pH from 4.1 to 6.3. Soils with coarser texture and/or higher base saturation will probably require lesser amounts of fly ash to attain the same final pH.

The increase in electrical conductivity of the soil saturation extract (EC_e) with application of fly ash varied among soils and crops (Table 33). The data, however, indicate substantial gains in the

TABLE 34 Effect of Fly Ash on Soil pH and on Calcium Plus Magnesium and Boron from Displaced Soil Solutions of Groseclose and Woodstown Soils Cropped to Corn for 2 Years[a]

Fly Ash Added[b] (metric tons/ha)	pH[c]	Ca + Mg[d] (meq/liter)	B[d] (μg/ml)
Groseclose Soil			
0	5.8	3.18	0.33
144	6.6	4.99	0.49
Woodstown Soil			
0	5.4	1.88	0.30
144	5.8	3.10	0.59

[a]Source: Plank *et al.* (1975).

[b]Weathered Glen Lyn fly ash; pH of a 1:1 water suspension of this fly ash was 7.8, and the material contained Ca + Mg and B at concentrations of 3.4 percent and 100 μg/g, respectively.

[c]The pH of the surface 15 cm of soil.

[d]Data shown are average concentrations of 4 consecutive depths, 0-15, 15-30, 30-46, and 46-61 cm.

content of total soluble salts (as measured by the electrical conductivity) of soils amended with fly ash. Similar observations were made by Mulford and Martens (1971), who noted that the EC_e of Tatum silt loam (Typic Hapludult, clayey, mixed, thermic) was increased from 1 mmho/cm to 4.4 mmhos/cm with application of 5 percent fly ash. Salt-sensitive crops are injured when the EC_e exceeds 4 mmhos/cm (Richards, 1954). Soils receiving high applications of fly ash might thus be unfavorably affected, and leaching of excess salts from the soil would become necessary for the maintenance of good crop growth.

Studies conducted by Page et al. (1977) on the efficiency of Colorado River water in leaching of soluble salts induced in Baywood sandy soil (Entic Haploxeroll, sandy, mixed, thermic) by the application of 5 percent fly ash to the top 3 cm of a soil column showed that approximately 60 surface cm of applied water were required to bring the salt level in the soil to levels comparable with the background levels. Achievement of such a quantity of water is feasible under normal practices in irrigation agriculture.

Application of 8 percent fly ash produced boron concentration in saturation extracts from the Arizo and Redding soils of approximately 8 and 11 μg/ml, respectively (Table 33). These concentrations are from fivefold to twentyfold that of the initial boron concentration. The hot-water-soluble boron (a diagnostic technique used to evaluate boron deficiency) of Groseclose and Woodstown soils (Table 34) and of Tatum soil (Mulford and Martens, 1971) was also increased on incorporation of fly ash into these soils. In the latter soil, application of 1.7 percent of fly ash (618 ppm total boron) increased the hot-water boron content of the soil from <1 to about 15 $\mu g/g$. These observations and others (Plank and Martens, 1974) show that soils could be greatly enriched in soluble boron by fly-ash application. As boron-sensitive crops are injured at relatively low concentrations in the saturation extract (1 μg of B/ml) (Bingham, 1973), leaching of excess boron in fly-ash-treated soils may be necessary for optimum growth of plants. Approximately the same amount of water that was required to leach excess salts in the top 3 cm of Baywood soil columns was also required to bring the boron concentration in leachates of this soil to levels comparable with the check soil (Page et al., 1977). This finding indicates that excess salinity and boron could simultaneously be reduced in soils amended with fly ash, if appropriate leaching regimes were maintained.

Application of 8 percent fly ash markedly increased the sulfur content of soils (Table 33). The 1N ammonium acetate (1N NH_4OAc) soluble-sulfur concentration in the soil at this rate was approximately 300 $\mu g/g$ SO_4-S. This concentration agrees well with predicted concentrations, based on the average SO_4-S of this fly ash (0.4 percent). Furthermore, addition of gypsum ($CaSO_4 \cdot 2H_2O$) to Arizo soil cropped with alfalfa produced concentrations of SO_4-S in the soil similar to those produced by fly ash when the latter was added at equal rates of sulfur (Elseewi et al., 1978b). The authors concluded that the availability of fly-ash sulfur appears equal to gypsum sulfur, an indication that sulfur in the former exists in a chemical form very similar to that of gypsum. Response of plants to sulfur in fly ash, as discussed in a subsequent section, was quite remarkable.

Three soils of varying chemical properties were incubated at moisture contents of one-third bar and saturation for a period of 29 weeks with rates of fly ash ranging from 0 to 1 percent (Page et al., 1977). Soil pH and salinity were increased regardless of the moisture content of incubation. No appreciable changes in the concentrations of iron, manganese, copper, zinc, molybdenum, nickel, and lead in the soil saturation extract were detected. However, concentrations of calcium, magnesium, barium, strontium, and boron in the soil saturation extract increased, in some cases rather dramatically, in relation to the amount of fly ash added. The increases varied in magnitude between soils, with acid soils, in general, showing the greatest increases. Diethylene-triaminepentaacetic-acid-(DTPA)-extractable metals also varied depending on rate of fly-ash addition and soil type. The incubation study showed that mixing the soil with up to 1 percent fly ash will not cause harmful enrichment of soil in trace metals, except possibly for boron.

Physical Properties of Fly-Ash-Amended Soils

Application of fly ash to agricultural soils in amounts greater than a few percent generally has a significant impact on the physical properties of soil, the quantities and effects being influenced by the properties of the specific fly ash and soil involved. Chang et al. (1977) conducted a greenhouse experiment with five soils amended with Mojave fly ash at rates ranging up to 50 percent by volume and cropped to barley. Physical parameters such as water content at 20 cbar, water release, bulk density, hydraulic conductivity, and modulus of rupture were measured. Results showed that the addition of up to 10 percent fly ash to the soil did not significantly affect the amount of water retained by soils. At rates greater than 10 percent by volume, water retention by soils increased with increasing rates of fly-ash application. The volume of water released to the plant root system was likewise unaffected by up to 10 percent fly ash in the soil but was significantly reduced at rates higher than 10 percent by volume. Addition of fly ash progressively decreased the bulk density of soils. Hydraulic conductivity measurements, however, indicated that although the bulk density was decreased, passage of water through soils reached a maximum at a fly-ash rate characteristic of each soil, beyond which it declined. Sharp reductions in the modulus of rupture (an index of soil strength) were obtained at low rates of fly-ash application to soils. This reduction would indicate less cohesiveness of soil particles after the addition of fly ash, thus increasing the potential of erosion of fly-ash-amended soils.

Mineral Composition of Plants Grown on Fly-Ash-Amended Soils

Effects of fly-ash incorporation in soils on the mineral composition of alfalfa grown on calcareous and acid soils amended with up to 8 percent fly ash are shown in Table 35. Plants were adequately fertilized with nitrogen, phosphorus, and potassium from a mixed fertilizer. Elements most affected by fly-ash applications are phosphorus, sulfur, calcium, sodium, zinc, boron, and manganese. The phosphorus and zinc contents were generally reduced with fly-ash application; those of calcium, sodium, and boron were generally increased. In alfalfa, as well as in several other plant species tested, manganese showed two distinctly different trends in plants grown on acid and on calcareous soils. Under acid-soil conditions, the first application of fly ash sharply reduces the manganese content of plants. Under calcareous conditions, plants show a moderate increase in manganese content up to a certain level of fly-ash addition (2 to 4 percent).

Reductions in phosphorus and zinc concentrations in the plant tissue were not sufficient to induce deficiencies of these elements in the plants. Reduced availability of zinc with application of alkaline fly ash was also observed by Schnappinger _et_ _al_. (1975) on corn grown on a slightly acid soil Frederick silt loam (Typic Paluedult, clayey, kaolinitic, mesic) amended with fly ash at rates ranging from 0.8 to 13 percent by weight. Addition of acidic fly ash to the same soil increased zinc uptake and corrected the deficiency in plants. Thus, although fly ash contains phosphorus and zinc, the high alkalinity of the material and of the soil and fly-ash mixtures appears to curtail the availability of these elements to plants.

The increase in sulfur availability to plants (Table 35) was associated with significant yield improvement in a number of plant species. Furthermore, the availability of fly-ash sulfur compares well with that of sulfuric acid, gypsum, and sewage sludge, as revealed by a series of greenhouse studies conducted by Elseewi _et_ _al_. (1978b,c) with alfalfa, Bermuda grass, white clover, and turnips. Results of the turnip experiment are shown in Table 36, as an example of the essentially equal availability of sulfur in fly ash, sewage sludge, and gypsum. Application of these materials at the rate of 25, 50, and 100

TABLE 35 Concentrations of Various Elements in Alfalfa Tops from Plants Grown on Fly-Ash-Amended Soils[a]

Fly Ash in Soil (percent)	Percent N	P	K	S	Ca	Mg	μg/g (dry weight) Na	Fe	B	Zn	Cu	Mn
Arizo Calcareous Soil												
0	2.2	0.47	2.9	0.08	1.4	0.22	317	85	106	39	5.9	73
1	2.7	0.27	1.9	0.19	2.0	0.20	318	80	148	23	5.3	144
2	2.7	0.27	1.9	0.11	2.4	0.24	550	85	183	25	6.1	148
4	3.0	0.30	2.1	0.19	3.2	0.30	593	105	222	35	5.8	141
8	3.0	0.27	1.8	0.23	2.5	0.32	853	105	337	28	6.0	92
Redding Acid Soil												
0	3.4	0.65	3.6	0.10	1.4	0.34	632	231	87	56	5.8	441
1	4.0	0.60	3.6	0.24	1.7	0.32	697	252	90	50	6.2	150
2	4.1	0.55	3.5	0.33	2.0	0.32	753	222	93	45	5.8	150
4	3.8	0.47	3.7	0.34	2.3	0.37	821	218	101	46	6.5	133
8	4.3	0.40	3.6	0.31	2.4	0.43	672	207	137	39	6.5	145

[a]Source: Page *et al.* (1977).

TABLE 36 Dry Matter Yield, Sulfur Content of Tissue, and Sulfur Uptake of Turnips Grown on Josephine Soil Amended with Three Rates of Fly Ash-Sulfur, Gypsum-Sulfur, and Sewage Sludge-Sulfur[a]

Rate of Sulfur Addition (mg/kg)	Dry Matter Yield (g/container)	Sulfur Content (percent, dry weight) Root	Petiole	Leaf	Sulfur Uptake (mg/container)
Check	25.8	0.02	0.01	0.02	3.9
25 gypsum	41.8	0.14	0.18	0.14	67.8
25 fly ash	42.3	0.16	0.14	0.11	58.3
25 sludge	39.8	0.13	0.13	0.06	40.6
50 gypsum	42.8	0.26	0.34	0.42	164.9
50 fly ash	45.8	0.24	0.27	0.21	101.3
50 sludge	53.8	0.15	0.22	0.17	88.6
100 gypsum	41.0	0.27	0.46	0.56	180.4
100 fly ash	48.8	0.28	0.36	0.31	143.5
100 sludge	64.5	0.22	0.39	0.42	199.8
LSD,[b] 5 percent	6.5	0.04	0.05	0.07	30.9

[a]Source: Elseewi *et al.* (1978b, c).

[b]Least significant difference.

mg of S/kg of soil increased the yield and sulfur content of turnips grown on the sulfur-deficient Josephine soil (Typic Haploxealts, fine, loamy, mixed, mesic).

Increased levels of sodium and boron in plants grown on fly-ash-amended soils (Table 35) had no adverse effects on the growth of alfalfa but were partially responsible for yield reductions in lettuce (Elseewi et al., 1978a). The availability of boron in fly ash to alfalfa was shown by Mulford and Martens (1971) and Plank and Martens (1974) to be essentially equal to that of sodium borate-boron; however, the availability decreased with time, as indicated by results of a 3-year field experiment with alfalfa (Plank and Martens, 1974).

Table 37 shows selected analyses of molybdenum in various plant species grown on calcareous and acid soils amended with fly ash. The molybdenum concentrations shown in Table 37 reached levels as high as 22.0 and 44.4 μg of Mo/g in alfalfa and white clover, respectively. The data, however, indicate that the availability of fly-ash molybdenum is inversely related to time. Doran and Martens (1972) observed that the availability of fly-ash molybdenum to alfalfa was approximately equal to that of sodium molybdate.

Analyses for strontium, barium, selenium, cobalt, scandium, rubidium, chromium, tungsten, antimony, silicon, calcium, cerium, thorium, beryllium, arsenic, samarium, europium, and mercury in plants grown on fly-ash-amended soils showed that only strontium, barium, selenium, cobalt, and cesium exhibit definite concentration trends in the plant tissue with increasing rates of fly-ash application to soil (Table 38). The increase in strontium may serve as an index of fly-ash deposition on soils and vegetation in areas adjacent to coal-fired power plants. The increase in selenium, when combined with the increase in molybdenum, indicates potential animal nutritional problems associated with fly-ash application. Selenium in the range 0.04 to 0.2 ppm and molybdenum at levels less than 1.0 ppm (dry weight basis) are required in animal diets (Allaway, 1968). At the same time, concentrations in plants in the ranges of 5 to 20 ppm of molybdenum and 4 to 5 ppm of selenium are potentially dangerous to animals (Allaway, 1968). Higher levels of molybdenum can produce a disorder called "molybdenosis," or molybdenum-induced copper deficiency in animals. Results from field and greenhouse studies with fly ash indicate that selenium concentrations in some crop species approached the foregoing critical levels (Furr et al., 1977; Straughan et al., 1979) and in one particular circumstance exceeded these levels (Gutenmann et al., 1976). Tests with small animals showed elevated selenium and molybdenum concentrations in their tissues and organs when they were fed plant rations grown on fly-ash-amended soils (Furr et al., 1978; Gutenmann et al., 1976; Stoewsand et al., 1978).

Plant Growth on Fly-Ash-Amended Soils

Rees and Sidrak (1956) noted that growth of Atriplex plants was improved with coal-ash application to soil in the presence of an adequate supply of nitrogen. Hodgson and Holliday (1966) grouped some crop species according to their tolerance to pulverized fuel ash (Table 39). Based on their classification, legumes, beets, and most grasses appear to be

TABLE 37 Molybdenum Concentrations of Various Plant Species Grown on Arizo Calcareous and Redding Acid Soils Amended with Fly Ash[a,b] (μg/g, dry weight)

Plant Species	Percentage of Fly Ash in Soil			
	0	2	4	8
Arizo Soil				
Alfalfa				
First clipping	4.5	7.5	17.0	22.0
Sixth clipping	1.4	2.7	4.3	9.5
Eleventh clipping	1.8	1.8	1.9	4.2
Bermuda grass				
First clipping	1.2	2.7	5.9	4.4
White clover				
Second clipping	2.5	1.7	10.0	17.5
Brittlebush				
Leaf sample	1.1	3.4	4.5	8.1
Redding Soil				
Alfalfa				
First clipping	0.9	5.2	7.8	17.2
Sixth clipping	2.0	3.4	7.5	9.5
Eleventh clipping	3.2	3.6	5.1	5.7
Bermuda grass				
First clipping	0.5	4.7	5.2	5.5
Fourth clipping	0.9	2.8	4.3	4.5
White clover				
Second clipping	2.8	11.6	44.4	41.6
Brittlebush				
Leaf sample	0.9	4.6	6.1	7.6

[a]Source: Page *et al.* (1979).

[b]Except for brittlebush, analysis was performed on composites of whole tops of plants.

tolerant of fly ash. Beans and peas, however, are sensitive exceptions. The authors indicated that response of plants to fly ash is determined mainly by their ability to withstand excessive concentrations of soluble salts, boron, and possibly high pH in the soil.

Application of fly ash, weathered for approximately 3 years in a power-plant disposal lagoon, at the rate of 144 metric tons/ha (6.4 percent) increased the yield of corn on Woodstown soil (Table 40) by approximately 27 percent (Plank et al., 1975). The increase in yield was attributed to increased water availability to plants. Application of 2.6 to 5.2 percent fly ash to Fredrick silt loam nearly doubled the yield of corn as a result of increased availability of zinc (Schnappinger et al., 1975). Increased rates of application of fly ash to Groseclose silty loam increased the availability of molybdenum to alfalfa and markedly increased its yield (Doran and Martens, 1972).

TABLE 38 Neutron Activation Analysis of Trace Elements in Plants Grown on Fly-Ash-Amended Arizo Soil Elements Showing Definite Concentration Trends[a] (μg/g, dry weight)

Fly Ash (percent)	Sr	Ba	Se	Co	Cs
Alfalfa					
0	30	4.5	0.2	0.14	0.026
0.5	77	9.3	1.1	0.12	0.060
1.0	125	18.0	1.7	0.12	0.071
2.0	196	25.0	2.8	0.16	0.070
4.0	226	28.0	4.5	0.36	0.053
8.0	364	45.0	4.6	0.45	0.105
Lettuce					
0	30	6.4	—	0.14	0.040
0.5	41	7.4	0.6	0.13	0.050
1.0	62	56.0	1.0	0.21	0.120
2.0	79	10.0	2.1	0.16	0.080
4.0	—	11.0	2.8	0.19	0.120
8.0	152	36.0	3.9	0.51	0.180

[a]Source: Page *et al.* (1977).

TABLE 39 Tolerance of Some Crop Species to Pulverized Fuel Ash[a]

Tolerant	Moderately Tolerant	Moderately Sensitive	Sensitive
Red beets	Cabbage	Brussels sprouts	Barley
Spinach	Turnip	Cauliflower	Oats
Sweet clover	Rape	Rye	Broad bean
Perennial ryegrass	White mustard	Wheat	Dwarf bean
Red fescue	Radish		Peas
	Carrot		Potatoes
	Red clover		Buckwheat
	White clover		Lettuce
	Alfalfa		Strawberry clover
	Meadow fescue		Meadow grass.

[a]Source: Hodgson and Holliday (1966).

TABLE 40 Yield Response of Corn and Alfalfa to Application of Fly Ash

Fly Ash Source	Rate (percent)	Yield Increase (percent)	Cause
Corn—Fredrick Silty Loam[a]			
Muskingum River	2.6	98.7	Increased availability of Zn
Muskingum River	5.2	98.7	
Phillip Sporn	3.4	17.3	
Phillip Sporn	6.8	48.0	
Corn—Westmoreland Silty Clay Loam[a]			
Muskingum River	7.9	43.8	Increased availability of Zn
Alfalfa—Groseclose Silty Loam[b]			
Kanawha River	2.2	25	Increased availability of Mo
Kanawha River	4.4	61	
Kanawha River	8.8	132	
Crawford Edison	0.3	61	
Crawford Edison	0.6	150	
Crawford Edison	1.2	132	
Corn—Woodstown Loamy Fine Sand[c]			
Glen Lyn	4.8	20	Increased availability of water
Glen Lyn	6.4	27	

[a] Schnappinger *et al.* (1975).
[b] Doran and Martens (1972).
[c] Plank *et al.* (1975).

Results of greenhouse experiments conducted by Page et al. (1977) with plants grown on calcareous and acid soils amended with Mojave fly ash are summarized in Table 41. Growth of alfalfa, Bermuda grass, white clover, Swiss chard, and brittlebush was improved by application of this fly ash to soils. The increase in yield was attributed, in all cases, to increased availability of sulfur to plants from fly ash. Yields of lettuce, corn, and radish were depressed as the rate of application of fly ash was increased. Yield depression in lettuce was significantly related to increased salinity and boron in the soil as a result of the fly ash.

Disposal in Surface Waters

The potential use of fly ash to purify polluted lake and river waters as well as ocean disposal has been studied by Tenney and Echelberger (1970), Higgins et al. (1976), and Thomson (1963). Fly ash has a strong

TABLE 41 Summary of Dry Matter Yield Data of Various Plant Species Grown under Greenhouse Conditions on Soils Amended with Variable Rates of Mojave Fly Ash[a]

Fly Ash in Soil (percent)	Relative Yield (percent)					
	Alfalfa	Bermuda Grass	White Clover	Lettuce	Swiss Chard	Brittle-bush[b]
Arizo Soil						
0	100	100	100	100	100	100
1	240	182	185	69	117	125
2	315	172	276	39	114	133
4	343	183	210	20	122	142
8	306	156	150	35	87	102
Redding Soil						
0	100	100	100	100	100	100
1	184	144	133	74	116	118
2	259	155	89	—	162	130
4	261	141	27	68	127	120
8	274	153	47	14	89	107
	Corn	Radish				
Domino Soil						
0	100	100				
2	109	53				
4	103	32				
8	52	21				
20	26	17				
40	4	17				
Redding Soil						
0	100	100				
2	47	50				
4	36	11				
8	27	13				
20	7	8				
40	7	13				

[a]Source: Page *et al.* (1977).

[b]A native desert plant species common to the Mojave Desert.

capacity to reduce excessive concentrations of phosphorus and also to remove undesirable color and organic contaminants from these waters. The reduction in phosphate concentrations is attributed to the formation of insoluble calcium-phosphate compounds under conditions of high pH and high content of free Ca^{2+} ions, which are prevalent in water extracts from many fly ashes. As much as 95 percent of the initial phosphorus concentration was removed by the addition of fly ash to Stone Lake (Michigan) water at the rate of 10 g/l (Tenney and Echelberger, 1970).

Higgins et al. (1976) studied the effectiveness of fly ash in removing excessive phosphorus concentrations from Lake Charles waters. They concluded that, although ponded fly ash is capable of removing some phosphorus, not enough is removed to justify its use for that purpose alone. They also showed that a 5-cm layer of fly ash on lake sediments was sufficient to prevent phosphorus release for extended periods.

Because of its relatively large surface area and residual carbon content, fly ash has a high adsorption capacity and is capable of removing many constituents from lake water. Examples of organic materials that were effectively removed by fly ash are alkylbenzenesulfonate (ABS), phenols, trinitrotoluene (TNT), refractory organic materials, colors, taste, odor, bacteria, and algae. Approximately 65 to 70 percent of the total organic materials in Stone Lake water were removed also by the addition of 10 g of fly ash/l to the lake water (Tenney and Echelberger, 1970).

Fly-ash additions to lake waters do not appear to impair other water-quality parameters. For example, water hardness and sulfate concentrations in lake waters did not increase beyond the critical levels prescribed by the U.S. Public Health Service (1962). Thomson (1963) has also shown that fly-ash addition at levels up to and including 10 g/l is not harmful to the aquatic life of a water body. He examined the mortality threshold of fish in fly-ash suspensions created by disposing of fly ash in the ocean and observed no injurious effects from the fly-ash concentrations indicated above.

Neutralization of Acid Mine Drainage

Various treatment and control procedures have been suggested for the reclamation of acid mine drainage (AMD) waters. Direct chemical neutralization of the acidic components of these waters appears to be the most widely accepted and used. AMD waters are characterized by low pH and high content of iron. Because of the presence of significant concentrations of calcium and OH ions in water extracts from most fly ashes, they have the potential of being used as agents in the neutralization of AMD waters. The effectiveness of fly ash in neutralizing AMD waters was demonstrated by Tenney and Echelberger (1970).

The amount of fly ash needed to increase the pH of these AMD waters to acceptable levels ($\simeq 7$) depends to a large extent on the acidity of the waters and the alkalinity of the fly ash used. Such AMD waters with high iron concentrations, hence higher acidities, tend to have higher buffering capacity than water containing lesser concentrations of iron and acidity, thus requiring greater quantities of fly ash to neutralize their acidity. Expressing total acidity in terms of milligrams of calcium carbonate per liter (mg of $CaCO_3$/l), Tenney and Echelberger (1970) have shown that waters containing approximately 3 g/l require fly-ash additions at the rate of 150 g/l to achieve neutrality (pH 7.3). Acid mine waters containing only about 100 mg/l total acidity required the addition of about 5 g of fly ash/l to achieve such neutrality. These results suggest that other acid industrial wastewaters could be treated with fly ash. Pickle liquor and plating solutions are examples. Additions of fly ash might also restore acid bog lakes.

Use of Fly Ash as a Physical Conditioner of Waste Sludges

A number of inert materials have been used to condition the sludge for dewatering. Examples of these materials are coke, bone ash, peat, paper pulp, ground blast-furnace slag, diatomaceous earth, sawdust, crushed coal, activated carbon, and fly ash.

Tenney and Echelberger (1970) indicated that fly ash is an excellent conditioner of various waste sludges such as those from paper mills, rubber plants, and municipal wastewater. They also conducted a study on the effectiveness of fly ash as a dewatering agent for activated sludge. Their results indicated that a specific fly-ash concentration of one part fly ash to ten parts sludge, by weight, gave optimum rate of dewatering. This rate corresponded to additions of 50 to 75 g of fly ash/l of sludge. Aside from improving the filterability of the sludge-fly-ash cake, fly ash lowered the residual moisture content of the cake and thereby increased its potential fuel value. Their study showed a reduction of 20 percent in the moisture content of the cake.

Tenney and Echelberger (1970) concluded that sludge particles are bonded to the surface of fly-ash particles by means of chemical and/or electrostatic interactions that give appreciable rigidity to the sludge-fly-ash lattice; therefore, the sludge particles are held securely during vacuum filtration. Such a mechanism increased porosity and allowed for the relatively unrestricted egress of water.

HEALTH EFFECTS

Occupational Health

Health effects from occupational exposure to trace metals from steam-electric plant processes have not been documented.

Public Health

The potential impact on the environment will come from stack emissions of fine particulates, which are subsequently deposited on the biological systems and from the disposal of collected residues. Studies conducted by Natusch and Wallace (1974) and Linton _et al_. (1976) indicate that a number of potentially toxic trace elements, including lead, cadmium, antimony, selenium, nickel, vanadium, zinc, cobalt, bromine, manganese, and SO_4, predominate in small particles emitted from high-temperature combustion sources. Deposition in the respiratory tract depends on particle size, as is illustrated in Figure 15. Particles of 1-μm diameter or less are deposited almost exclusively in the lung (Figure 15). As was shown earlier, such particles are enriched in trace elements to a great extent, particularly at the surface that comes in contact with tissues.

Little is known about the fate of metals emitted from tall stacks. Both modeling efforts and systematic grid sampling around power plants have indicated that less than 10 percent of the metals emissions can be accounted for within 50 km of the plant. The predicted air concentrations of known mutagens, teratogens, and carcinogens near a

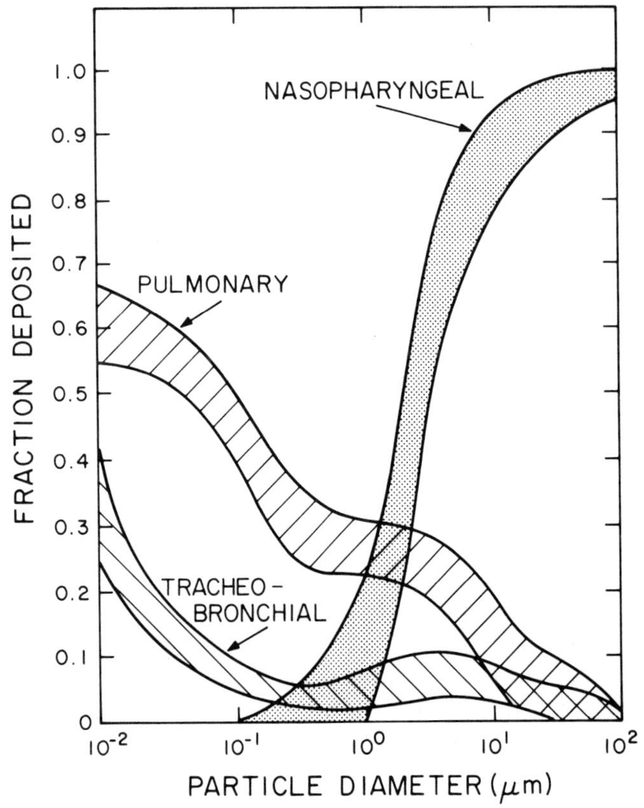

FIGURE 15 Respiratory deposition efficiencies for inhaled particles (U.S. Department of Health, Education, and Welfare, 1969).

1400-MWe plant are (except for beryllium) orders of magnitude less than acceptable concentrations (Table 42) (Crockett and Kinnison, 1977; Swaine, 1977; Vaughan et al., 1975).

The specific trace-metal components of the particulate portion of the SO_x-particulate complex strongly influence the response of experimental animals to the aerosol mixture. Copper, manganese, vanadium, and iron particles plus SO_2 produce increases in airway resistance in animals several times greater than that produced by SO_2 alone. These metals are known promoting agents of the conversion of sulfur dioxide to sulfuric acid, considered to be the most irritating sulfate species (Amdur, 1976). In experimental heterogeneous droplet systems, manganese and iron are cocatalysts of the conversion of sulfur dioxide to sulfate (Beilke and Gravenhorst, 1977). Trace-metal synergism with other air pollutants may well be the mechanism of greatest concern to public health, especially considering the long-range transport undergone by gases and small particulates emitted from tall stacks.

Vaughan et al. (1975) modeled the behavior and accumulation of trace metals in soil, water, biota, and humans that could be emitted from a hypothetical 1400-MWe coal-fired plant sited in a semiarid western watershed and burning representative western coal. They concluded that the hazard incurred from direct inhalation would be negligible, because respirable (submicrometer) particulates constituted 1/100 or less of the total particle burden, which was an order of magnitude less than present urban air-pollution concentrations. This finding was supported by Baser and Morris (1977), who computed excess deaths attributable to trace

TABLE 42 Predicted Air Concentrations of Airborne Trace Elements in the Vicinity of a 1400-MWe Coal-Fired Power Plant Compared with Acceptable Air Concentrations[a]

Element	Predicted Air Concentrations ($\mu g\ m^{-3}$)	Acceptable Air Concentrations ($\mu g\ m^{-3}$)
As (V) or total	1.2×10^{-4}	0.1
As (III)	—	1×10^{-5}
Be	2.9×10^{-4}	5×10^{-3}
Cd	1.2×10^{-4}	0.05
Cr	7.2×10^{-6}	0.05
Cr (VI insol)	—	1×10^{-6}
F	2×10^{-4}	0.01
Hg	1.2×10^{-3}	0.1
Hg organic	—	0.01
Mo	2.4×10^{-4}	0.1
Ni	4.8×10^{-4}	0.01
Ni carbonyl	—	1×10^{-6}
Pb	2.4×10^{-4}	1.0
Sb	2.4×10^{-5}	0.1
Se	2.4×10^{-5}	0.1
Ti	3.0×10^{-6}	0.01
U	3.0×10^{-3}	0.01
V	2.4×10^{-4}	0.05

[a] Source: Van Hook and Shults (1976).

metal exposure from a hypothetical 1000-MWe coal-burning plant (Table 43). The predicted impacts on drinking-water concentrations were several orders of magnitude lower than those proposed as maximum acceptable concentration standards for potability. Cadmium, molybdenum, and tungsten showed 30 to 100 percent increases in green plants, assuming partial solubility of exogenously supplied metals. The incremental burden for the remaining metals was about 11 percent of reported content. The size of these changes, which occur over 40 years of plant operation, are small in comparison with those routinely encountered through fertilization of crops. Attention was called to molybdenum, which has a stimulating effect on both anaerobic denitrification and nitrogen fixation reactions. The first reaction could be destructive to agriculture by depleting soils of available nitrogen; the second is a limiting one and critical to agricultural productivity. The study concludes that the ingestion of plants for food appears to be the only consequential pathway of anthropogenic elements to man. In general, excessive bioaccumulation is not expected unless the bioavailability of exogenous fractions is orders of magnitude greater than that of endogenous fractions.

TABLE 43 Projected Annual Excess Mortality from Selected Trace Metals and Iron Related to Air Emissions from a 1000-MWe Coal-Fired Plant[a]

Trace Metal	Range of Conc in Coal (ppm)	Excess Mortality Coefficient[b]	Range of Magnitude of Annual Excess Mortality (exponents)
Cd	0.01-28	15.1	E-4 to E-1
Cr	4.4-60	9.9	E-3 to E-2
V	10-50	11.2	E-2 to E-2
Fe	1500-7520	15.8	E0 to E0
Mn	6-100	3.1	E-4 to E-2
Ni	4-61	5.3	E-3 to E-2
Zn	5.3-1610	3.7	E-3 to E-1
Hg	0.04-33	[c]	E-4 to E-1
Pb	2-106	[c]	E-3 to E-1
Be	0.1-1000	[c]	E-5 to E-1
As	2-25	[c]	E-3 to E-1
Se	1.2-7.7	[c]	E-3 to E-2

[a]Source: Baser and Morris (1977).

[b]Unit of coefficient is excess deaths per 10^5 persons per microgram of metal per cubic meter of air.

[c]Coefficient not given; 15.1 used.

Radioactive elements are present in stack gas and ash piles. ^{222}Rn and ^{220}Rn are released in gases from the stacks and account for most of the 7×10^{-3} mrem/yr per 1400-MWe coal-fired plant. In comparison, the whole-body dose equivalent rate from all natural sources is 80 ± 40 mrem/yr. Assuming that the radon-release percentage from coal-ash piles is similar to the radon-release percentage from uranium-tailings piles (about 5 percent), ash collected and sorted over the 30-year lifetime of a 1000-MWe plant might have a 3.4- to 15-times greater radon-release rate, depending on the type of coal, than natural background (Van Hook and Shults, 1976).

Ponding and ground burial of sludge, slag, and fly ash represent probable sources of water contamination. These residues contain the less-volatile trace elements, but acidification of fly ash increases the solubility of most elements, an indicator of biological availability for leaching. Arsenic, lead, and selenium are present in the scrubber liquor in concentrations greater than the allowable drinking-water standards (Bern, 1976; U.S. Public Health Service, 1962). One test project showed no effects on groundwater from disposal of either treated or untreated sludge (Fling, 1976), but it is difficult to extrapolate these findings to other locations and soils because of the wide variability in soils for filtration. Measures to prevent contamination of groundwater and surface waters may be required (Bern, 1976).

6

SYNTHETIC-FUEL PROCESSES (LIQUEFACTION, GASIFICATION) AND COKE PRODUCTION FROM COAL

Through thermal processing, coal may be converted from a solid to hydrocarbon liquids and gases, which are cleaner and more conveniently handled. These products are referred to as synthetic fuels (or synfuels). Coal can also be processed thermally to produce a solid, high-carbon coke used commercially to reduce iron ore to metallic iron. Because the coking operation is basically similar to proposed pyrolysis processes for synfuels production, it is also dealt with in this chapter.

Although commercial coking processes have operated for almost 100 years, little information about the fate of trace elements in coking operations has been made available to the public. No large commercial synthetic fuel plants are operating within the United States, and only a few are operating elsewhere. There is little available information about the fate of trace elements in such processes. Within the United States, however, there are many research programs devoted to the development of synfuel processes, and from some of these programs a small amount of data has begun to emerge. This chapter discusses the unit operations characteristic of the synfuel processes (some of which are similar to those in the coking processes) and their potential for concentrating or emitting trace elements. In addition, the sparse data that are available have been summarized or referenced.

SYNTHETIC-FUEL PROCESSES

When coals are heated in the virtual absence of free oxygen to about 400°C and above, they pyrolyze or decompose. Lower-molecular-weight substances evolve as volatile matter, leaving a residual char or coke. The carbon-rich char or coke may be used as a fuel for combustion or as a carbon source for steam gasification ($C + H_2O + heat \rightarrow CO + H_2$) or as a reducing agent for the production of iron from its ore. A portion of the volatile matter can be condensed and recovered as liquid tars or oils. If the coal is slurried in a hydrocarbon oil and then heated under pressure, the same basic chemical reactions take place. However, in this process, referred to as liquefaction, substantially more of the coal is converted to liquids that may then be separated from any unliquefied coal.

These operations are shown schematically in Figure 16. The pyrolysis process operated at 400-500°C can be considered a means principally for obtaining liquids from coal (although a substantial proportion of residual char is also recovered); when carried out at about 1000°C with a premium coking coal, its principal purpose is to produce coke for the smelting of iron ore in a blast furnace (although some tar by-products are also recovered). Coal or char may be fed to the gasification process; with coal, the volatile products would be converted to gas (though condensible liquids could be recovered depending on the reactor design). The undissolved coal from the liquefaction process is normally ash rich; the organic content of this fraction depends on the severity of the liquefaction process and the procedure used to separate the liquids from the residue. This residue can be pyrolyzed to recover additional liquids, it can be burned as fuel, or it can be disposed of. From all the processes, there is normally an aqueous liquor produced by condensing steam that is formed in the process.

Figure 17 is a schematic flow plan showing the major unit operations in a typical coal liquefaction plant. The process includes a gasifier to convert carbon from unliquefied coal to gas; by following the coal preparation flow and the flow of streams into and out of the gasifier block, the flow plan for a "stand alone" gasifier plant can be envisioned. The coker in Figure 17 can be regarded as a pyrolysis reactor; thus, from the flow through this system, a pyrolysis plant or coking operation can be inferred.

Coking and synfuel processes (i.e., pyrolysis, gasification, and liquefaction) share a number of common unit operations that could concentrate or evolve trace elements. These unit operations are shown in Figure 18 and are listed in Table 44. Commonly, a gasification plant will produce tars as a by-product, and invariably a liquefaction plant will contain a gasifier to produce hydrogen or fuel gas. In addition to the operations unique to coking and synfuels production, coal storage and handling processes and furnace-flue-gas emissions are similar to those discussed in the sections of the report for utility plants.

SYNTHETIC-FUEL EFFLUENT STREAMS

Commercial slot-type coke ovens emit volatile pyrolysis products (tars, gases) during charging, through leaks in the oven, and during coke discharge and quenching operations. The fate of trace elements during these phases of operation is unknown. In contrast, synthetic-fuel processing occurs in closed and sealed systems, as in petroleum refinery operations. However, as with any process, there are specific effluent streams. The effluent streams in which trace elements could be emitted to the environment are shown in Figure 18 and are listed in Table 45. Whether the trace-element content of any particular stream represents a potential hazard to the environment depends on three factors: the trace-element content of the prepared feed coal, the specific processing conditions, such as temperature and pressure (affecting the conversion

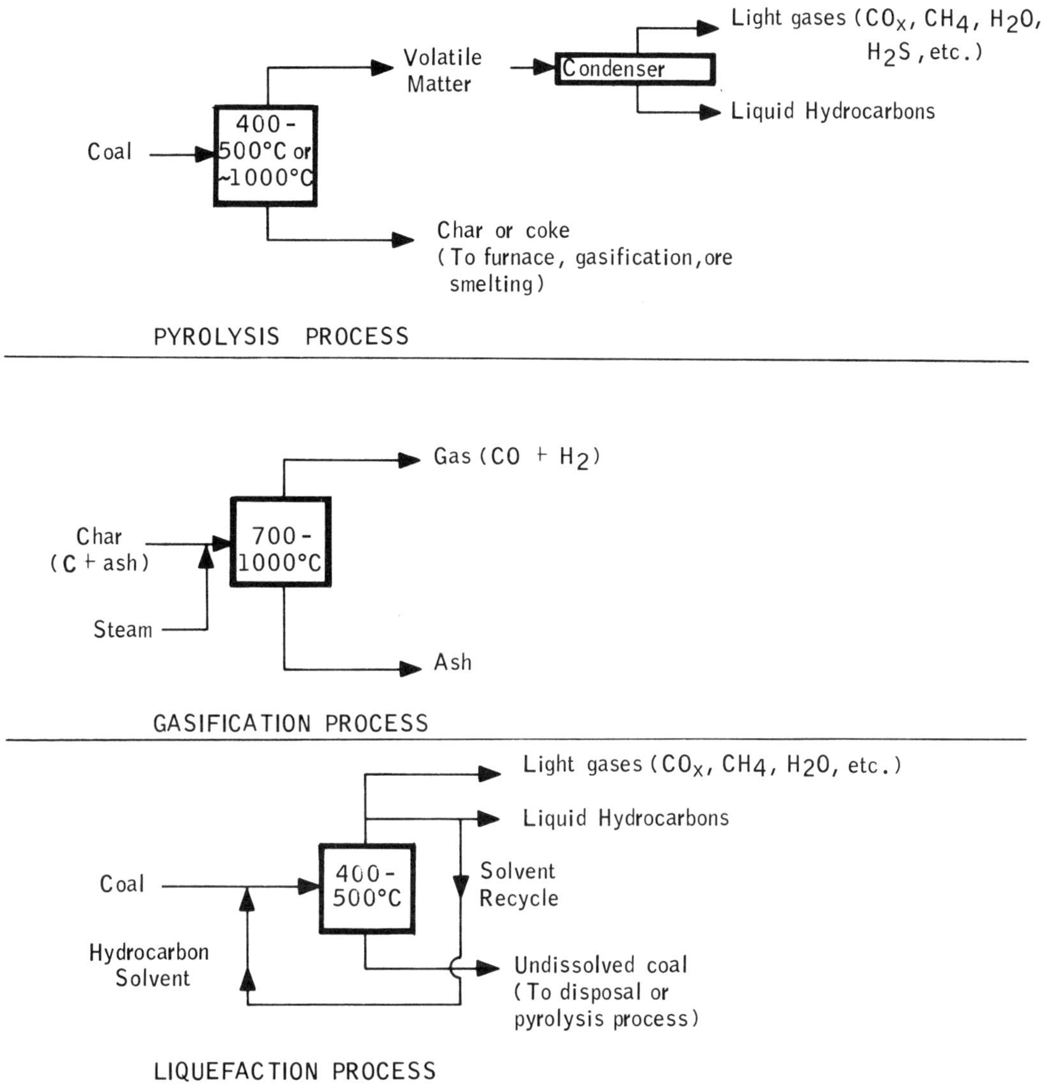

FIGURE 16 Schematic description of principal coal-conversion processes.

of trace elements to labile forms), and the control measures exerted to reduce emissions.

Effluent streams that have the potential for being especially troublesome and that are unique to synfuel processes are the following:

• The quenching and disposal of gasifier ash. Conceivably, the chemical forms of trace elements in gasifier ashes could be different from those in furnace ashes, because the gasifier atmosphere is reducing whereas the furnace atmosphere is oxidizing and is generally at a higher temperature. No data bearing on this question are available.

• Gas scrubber water and condensate. Sparse data that are available (Forney *et al.*, 1975; Sharkey *et al.*, 1975) indicate that treated product gases may be essentially free of metallic elements. Normally,

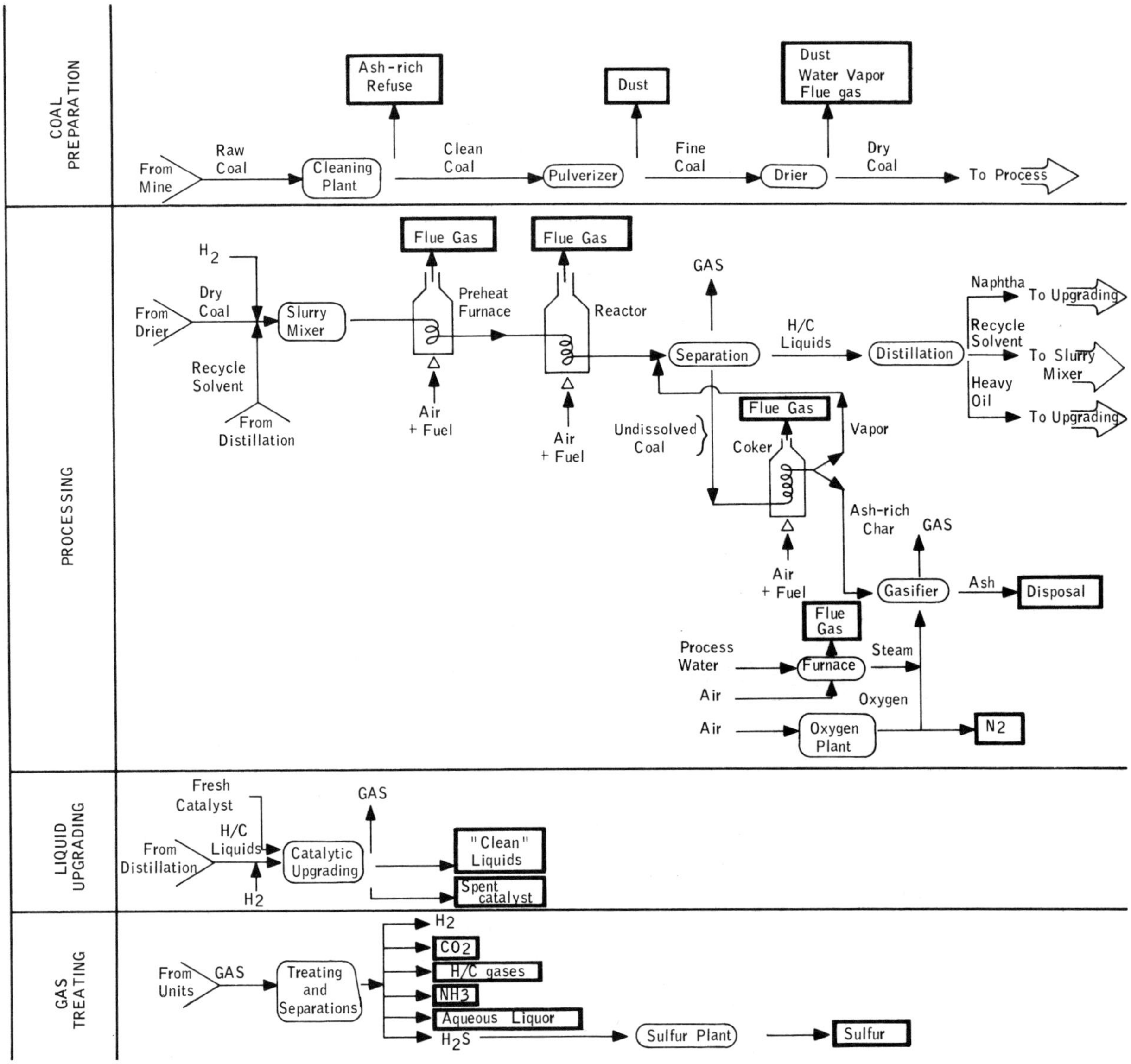

FIGURE 17 Flow plan showing major unit operations in a typical coal liquefaction plant. Effluent streams in which trace elements might leave the plant are indicated by heavy-lined boxes.

those trace elements that are volatilized in the gasifier (possibly cadmium, chlorine, fluorine, mercury, sulfur, selenium, tellurium, and zinc) could be expected to be trapped in the gas condensate liquor and the acid-gas removal system. However, until current research develops appropriate methods of treatment, inadequately treated gases that are burned hot (i.e., in an adjacent utility plant) could contain volatile elements that would be emitted to the atmosphere. The aqueous condensate liquor from coking plants probably would be similar to gasifier liquors. Treatment of this stream will be critical to the quality of effluent waters.

• Acid-gas removal system. Amine or carbonate systems to remove acid gases might collect trace elements not condensed in the aqueous

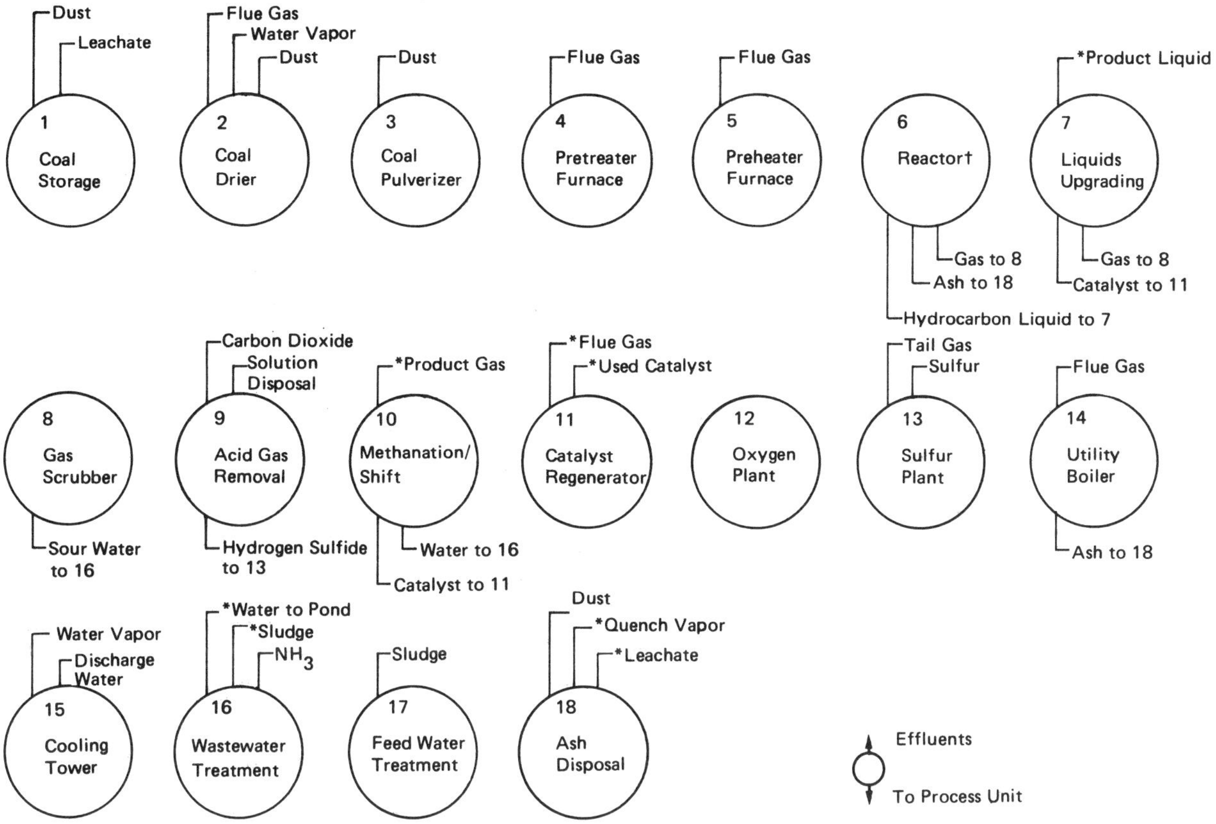

*Streams likely to be unique to synfuels processes compared to similar petroleum operations.
†Reactor (Item 6) may be liquefaction reactor, gasification reactor, or pyrolysis (coking) reactor and associated separations equipment.

FIGURE 18 Unit operations in liquefaction and gasification plants. The applicability of specific unit operations to specific processes is given in Table 44.

liquor, which could result in disposal problems for this stream. Although there are no data available to indicate that such a phenomenon would occur, power-plant scrubbers using lime or limestone to remove SO_2 do not appear to trap noxious trace elements (V. E. Swanson, Geological Society of America, Boulder, Colorado, personal communication, 1977).

• Tail gas from sulfur plant. All synfuel processes are likely to produce sulfur as a by-product. Though this is common practice in petroleum refineries, the feed-trace-element contents differ. Coal trace elements might produce unique problems in sulfur plants associated with synfuel installations because of the significantly greater amounts of trace elements in coal than in petroleum; though here also, no data are available.

• Catalyst disposal. Catalysts employed to upgrade coal liquids will likely concentrate some trace elements. Disposal of used catalysts could present unique problems related to the fate of the trace elements.

• Product liquid utilization. Many trace elements indigenous to the feed coal will be found in product hydrocarbon liquids from coal liquefaction. Their ultimate fate will depend on the final use of the

TABLE 44 Unit Operations in Coking and Synfuels Plants

Operation	Coke Plant	Gasification[a]	Liquefaction[b]
1. Coal storage	X	X	X
2. Coal drier	—	X	X
3. Coal pulverizer	X	X	X
4. Pretreater furnace	—	S	—
5. Preheater furnace	—	—	X
6. Reactor	X	X	X
7. Liquids upgrading	X	S	X
8. Gas scrubber	X	X	G
9. Acid gas removal	X	X	G
10. Methanation/shift	—	X	—
11. Catalyst regenerator	—	S	X
12. Oxygen plant	—	S	—
13. Sulfur plant	—	X	X
14. Utility boiler	—	X	X
15. Cooling tower	—	X	X
16. Wastewater treatment	X	X	X
17. Feed-water treatment	X	X	X
18. Ash disposal	—	X	G

[a]S, some plants.

[b]G, part of gasification plant to supply hydrogen for liquefaction process.

liquid, which is a legitimate concern in any consideration of synfuel processes.

No published data have been found for trace-element contents of coking plant streams, though some information may exist in company files and government agencies concerned with occupational safety. Only a limited number of references to trace elements in synfuel processes could be found. Table 46 presents a list summarizing synfuel-related trace-element data sources; a summary of the data for the elements is presented in Tables 47 through 51.

Other than the references listed in Table 46, no publications, of the many addressing trace elements in synfuel processes, report specific data. However, general discussions of the subject from the viewpoints of numerous authorities may be of interest. Therefore, those publications that have been consulted in the preparation of this chapter but that are not specifically cited are listed in the bibliography at the end of the chapter.

TABLE 45 Unit Operation Effluents and Products in Most Coking and Synfuels Processes

Effluents and Products	From Unit No.[a]
Dust	1, 2, 3, 18
Leachate	1, 18
Water vapor	2, 15
Flue gas	2, 4, 5, 11, 14
Hydrocarbon liquids	7
Carbon dioxide	9
Product gas	10
Used catalyst	11
Tail gas	13
Sulfur	13
Cooling-tower discharge water	15
Ammonia	16
Sludge	16, 17
Water (treated) to pond	16
Quench vapor	18

[a]As listed in Table 44.

The only reference encountered that discusses methods for dealing with trace-element effluents is an Environmental Impact Statement for a proposed gasification facility (U.S. Bureau of Reclamation, 1977). However, because no data are supplied, it is unclear whether the procedures will be applicable.

Few generalizations about the fate of trace elements in the many synfuel processes being investigated can be provided from available published data. Especially critical is the inability to predict how selected trace elements are likely to partition into specific streams in any process. Research programs devoted to the establishment of generalizable data are required. The major areas that should be investigated, using a variety of coals containing various concentrations of trace elements, are the following:

- The effect of process variables (atmosphere, temperature, pressure) in gasification on the partitioning of trace elements among the gas stream, condensates, and the residual ash.
  - Methods for treating condensate and scrubber streams from gasification processes.
  - The leaching of trace elements from disposed gasifier residues.
- The partitioning of trace elements between products and residues in liquefaction and pyrolysis processes.

TABLE 46 Sources of Trace-Element Data Pertaining to the Synfuels Process

Table	Reference	Process Stream
Liquefaction		
47	Coleman *et al.* (1976)	Concentration of trace element in solvent refined coal (SRC)
47	Jahnig and Magee (1975)	Concentration of trace element in solvent refined coal (SRC)
47	Schultz *et al.* (1977)	Concentration of trace element in synthoil
47	Given *et al.* (1975)	Concentration of trace element in coal-derived oils
48	Griffin *et al.* (1978)	Concentration of trace element in H-coal bottoms
48	Jahnig and Magee (1975)	Percentage of feed elements in SRC
48	Schultz *et al.* (1977)	Percentage of feed elements in product oil, centrifuge residue, and scrubber effluent
Gasification		
49	Sather *et al.* (1975)	Concentration of trace element in Lurgi residue ("ash")
49	Griffin *et al.* (1978)	Concentration of trace element in Lurgi residue ("ash")
49	Sather *et al.* (1975)	Loss of trace element through Lurgi gasifier
49	Attari *et al.* (1975)	Loss of trace element through hygas gasifier
50	Sharkey *et al.* (1975)	Concentration of trace element in synthane condensate
50	Baria (1975)	Percentage of feed elements in synthane condensate
50	Forney *et al.* (1975)	Percentage of feed elements in synthane condensate, tar, char, and fines
50	Somerville *et al.* (1977)	Percentage of feed elements in Lurgi ash, oil, tar, and liquor

Table 52 shows the scale of the process that would likely be necessary to investigate the areas mentioned above. Relatively small-scale batch processes would serve in some instances, whereas continuous process units would be required in others. Some data applicable to synfuel processes could be obtained by studying process streams from commercial coke plants.

HEALTH EFFECTS

Coking

Most of the nation's coke is produced in Northern Appalachia and the Central regions (Appendix A; Drysdale and Calef, 1977). Aside from

TABLE 47 Synfuels Trace-Element Concentrations in Solvent Refined Coal, Synthoil, and Coal-Derived Oils

Element	Process: Stream: Coal: Concentration: Reference:	Solvent Refined Coal (SRC) Product SRC Pittsburgh 8 ppm Coleman *et al.* (1976)	SRC Product SRC Unspecified ppm Jahnig and Magee (1975)	Synthoil Product Oil Western Kentucky Homestead ppm Schultz *et al.* (1977)	Unspecified Bench-Scale Process Product Oils Utah ppm Given *et al.* (1975)	Illinois 6	Kentucky
Antimony		—	0.3	—	—	—	—
Arsenic		—	0.5	—	—	—	—
Beryllium		—	0.4	—	4×10^{-4}	0.01	9×10^{-4}
Boron		—	51	—	0.15	0.12	0.03
Cadmium		<0.07	<0.1	0.8	—	—	—
Chromium		5.9	0.9	7.6	0.05	0.12	0.05
Copper		<9.5	2.5	2.7	0.02	0.03	0.01
Iron		423	161	—	—	—	—
Lead		<0.5	0.4	1.1	—	—	—
Manganese		21.6	1	11	0.05	0.12	0.05
Mercury		—	0.01	—	—	—	—
Nickel		23	4	6.6	0.15	0.08	0.03
Selenium		—	<1	—	—	—	—
Tin		—	9	—	—	—	—
Vanadium		—	16	—	0.06	0.25	0.004
Zinc		7.6	3	—	0.12	0.12	0.05

TABLE 48 Synfuels Trace-Element Concentrations and Percentage of Feed Elements in Other Liquefaction Processes

Element	Process: Stream: Coal: Concentration: Reference:	H-Coal Vacuum Bottoms Illinois Log_{10} ppm Griffin *et al.* (1978)	Solvent Refined Coal Product Unspecified Percent of Feed Element in Stream[a] Jahnig and Magee (1975)	Synthoil Product Western Kentucky Homestead Percent of Feed Element in Stream Schultz *et al.* (1977)	Synthoil Centrifuge Residue Western Kentucky Homestead Percent of Feed Element in Stream Schultz *et al.* (1977)	Synthoil Effluent Western Kentucky Homestead Percent of Feed Element in Stream Schultz *et al.* (1977)
Antimony		<1	2	—	—	—
Arsenic		1-2	4	—	—	—
Beryllium		1-2	58	—	—	—
Boron		3-4	80	—	—	—
Cadmium		<1	5	37	65	0.3
Chromium		2-3	2	48	75	1
Copper		1-2	78	31	74	2
Iron		>4	<1	—	—	—
Lead		1-2	3	33	77	3
Manganese		2-3	4	30	70	0.03
Mercury		<1	75	—	—	—
Nickel		1-2	6	63	72	0.2
Selenium		1-2	21	—	—	—
Tin		<1	14	—	—	—
Vanadium		2-3	7	—	—	—
Zinc		2-3	12	—	—	—

[a]Based on assumed yield of 75 percent (solvent refined coal), dry coal basis.

TABLE 49 Synfuels Trace-Element Concentrations in Lurgi Residue and Losses Through Lurgi and Hygas Gasifiers

Element	Process:	Lurgi		Lurgi	Lurgi		Hygas
	Stream:	Residue ("ash")		Residue ("ash")	Loss Through Gasifier		Loss Through Gasifier
	Coal:	Illinois 5	Illinois 6	Unspecified	Illinois 5[a]	Illinois 6[a]	Unspecified
	Concentration:	ppm		ppm	Weight Percent[b]		Weight Percent
	Reference:	Sather *et al.* (1975)		Griffin *et al.* (1978)	Sather *et al.* (1975)[a]		Attari *et al.* (1975)
Antimony		0.3	0.2	1-10	91	81	33
Arsenic		0.3	0.1	1-10	98	99	65
Beryllium		21	13.7	10-100	6	—	18
Boron		673	622	100-1000	79	—	—
Cadmium		<0.3	<0.3	<1	—	—	62
Chromium		570	755	100-1000	0	0	0
Copper		273	239	10-100	0	0	—
Iron		15×10^4	5.2×10^4	10^4	0	0	—
Lead		210	71	10-100	32	—	63
Manganese		320	200	>1000	0	0	—
Mercury		0.01	0.02	<1	99.5	99.6	96
Nickel		462	456	10-100	0	0	24
Selenium		—	—	<1	—	—	74
Tin		—	—	—	—	—	—
Vanadium		181	301	100-1000	19	0	30
Zinc		15.8	4.7	100-1000	29	0	—

[a]Based on residue representing 9.43 percent and 11.1 percent of dry feed coal (from ash balance calculation).

[b]Weight percentage of trace elements in feed not found in ash-rich residue.

TABLE 50 Synfuels Trace-Element Concentrations in Synthane Condensate and Percentage of Feed Elements in Related Gasification Processes

Element	Process: / Stream: / Coal: / Concentration: / Reference:	Synthane Condensate Illinois 6 ppb Sharkey *et al.* (1975)	Synthane Condensate — Percent of Feed Element in Stream Baria (1975)	Synthane Condensate —	Tar —	Char —	Fines —
				Percent of Feed Element in Stream Forney *et al.* (1975)[a]			
Antimony		—	—	0	0	203.1	2.0
Arsenic		30	0.06	0.2	3.0	116	4.3
Beryllium		—	—	0	0.2	138.4	4.2
Boron		—	—	75.7	0.9	95.1	0.6
Cadmium		—	—	0	1.4	404.1	10.4
Chromium		6	0.02	0	0.2	48	0.4
Copper		20	0.09	0	0.3	29.8	1.1
Iron		3×10^7	—				
Lead		—	—	0.1	1.2	656.4	5.1
Manganese		40	0.21	0.1	0.1	43.7	0.0
Mercury		—	—	39.9	19.6	15.7	1.2
Nickel		30	0.13	0.1	0.3	55.7	1.0
Selenium		360	9.23	11.1	0.5	310.5	9.9
Tin		20	0.70	0	0.2	68.9	2.6
Vanadium		3	0.005	0	0.1	53.4	0.9
Zinc		60	0.1	0.1	0.1	71.9	0.6

[a]Averages of three runs. Elemental balances range from 48 to 660 percent.

TABLE 51 Distribution of Selected Inorganic Trace Elements from Mercer County Coal, Indianhead Mine, North Dakota, During Sasol Gasification Test of a Lurgi Gasifier[a]

Element	Parts per Million in Coal	Percentage of Feed Element in Stream[b,c]			
		Gasifier Ash	Oil	Tar	Liquor
Sb	0.3	100	—	—	—
As	8	90.8	3.2	4.3	1.7
Be	0.3	99.1	—	0.9	—
B	56	99.3	0.03	0.03	0.7
Cd	<1	14.5	1.6	4.5	79.4
Cr	5.3	97.7	0.3	1.8	0.2
Cu	10.6	93.5	0.7	5.3	0.5
Fe	7936	99.0	0.02	0.95	0.01
Pb	2.7	88.3	0.7	10.9	0.1
Mn	70.7	99.2	<0.01	0.7	0.1
Hg	0.2	1.4	0.5	38.2	59.9
Ni	6.7	93.3	0.7	5.7	0.3
Se	0.4	13.8	1.3	2.8	1.5
Sn	0.27	100	—	—	—
V	21.3	99.0	0.01	1.0	0.01
Zn	6.7	74.7	1.6	3.4	20.3

[a]Source: Somerville *et al.* (1977).

[b]All elemental balances normalized to 100 percent total except selenium, which was assumed to exit with gas.

[c]Dashes indicate no data.

TABLE 52 Potential Sources for Data on Trace Elements in Synfuels and Coking Process

Process	Bench Batch	Bench Continuous	Pilot or Commercial
Pyrolysis partitioning	X	—	Coke plant
Liquefaction partitioning	—	X	X
Gasification, gas versus residue	X	—	—
Scrubber solutions	?	X	Coke plant
Gasification ash leach	X	—	—

total particulates, inorganic emissions from coking are not well characterized, but synergistic interaction of metals with other pollutants could contribute to effects on population near the plant.

As an illustration of the potential impact on local air quality, the Clairton Coke Works, largest in the country, emits from approximately 25 tons of particulates per day (Allegheny County Bureau of Air Pollution Control, 1977; Kenson et al., 1976).

The elevated lung-cancer mortality rate experienced by coke workers is most likely attributable to organics and major pollutants (Lakowicz et al., 1978; Laskin et al., 1975), although metals could play a role in potentiating and enhancing the effects of gases, organics, and particulates that workers are exposed to during the coking process.

A study of 3305 workers and 6475 controls revealed 2.5 times the lung-cancer mortality expected. There was also gradation of risk with type and duration of exposure. The relative risks were 1.70 for all men with less than 5 years' experience at the start of follow-up; 2.10 for men with 5 or more years side-oven experience only; 3.22 for men with 5 or more years of mixed topside and side-oven experience; and 6.87 for men with 5 or more years employed at full topside (Redmond et al., 1972).

Synthetic-Fuel Plants

Potential effects of trace elements in synfuel plants may be considered in two categories, public health (environmental) effects and occupational health effects.

Factors that could affect the routine release of trace elements to the environment were discussed in the preceding paragraphs. These may be categorized as coal-dust emissions, flue-gas emissions, and water contamination (from ash leachate and process condensate). Except for condensate water, which can be adequately treated before discharge, the potential public health effects in these three areas are expected to be only minimally different in kind (as already discussed) from those experienced as similar emissions from utility power plants. Synthetic-fuel plants will probably be quite large, consuming coal at rates equivalent to or larger than those of very large power plants. Therefore any problem related to ash disposal in synfuels plants would be at least equivalent in magnitude to the problems experienced at power plants. However, only about 10 percent of the coal consumed would be burned in a synthetic-fuel plant, meaning that flue-gas emissions would be significantly lower per ton of coal consumed in synfuels plants in comparison with large utility power plants.

The industrial hygiene aspects of synfuel plants will, however, differ from those indigenous to utility power plants. As has been discussed, processing is accomplished in closed systems, so that routine contact with concentrated or labile forms of trace elements will be negligible. System leaks should be minimal (comparable with experience in well-run petroleum refineries). However, because of the higher concentration of trace elements in the coal-derived synfuel in-plant process streams, any system leaks would have the potential for emitting larger amounts of trace elements than would be emitted from petroleum refineries. The concentration in some selected streams would also require that precautions be exercised during periods of maintenance when the systems are opened. However, it is likely that such precautions will be exercised because of the carcinogenicity of polynuclear aromatics that are found in coal-derived hydrocarbon liquids.

Distribution and utilization of the synthetic fuels should also be a public health concern. Coal-derived gases will be unlikely to contain any harmful trace elements because of the procedures used to clean product gases. Hydrocarbon liquids (asphalts, fuel oils, gasolines) might contain troublesome trace elements depending on the extent of processing to which they are subjected. Because these liquids are highly aromatic, contact by workers and consumers should be minimized anyway; therefore trace-element concerns are really secondary. Combustion of the liquids would likely result in emission of contained trace elements. This matter should be considered if coal-derived liquids (except those produced by reforming gases) are produced commercially.

BIBLIOGRAPHY

This bibliography represents a thorough literature search on synthetic-fuel processes and coke production from coal.

Attari, A. 1973. Fate of trace constituents of coal during gasification. EPA-650/2-73-004. U.S. Environmental Protection Agency. 31 pp.

Attari, A., M. Mensinger, and J.C. Pau. 1975. Fate of trace constituents in coal gasification. EPA-659/2-73-004. U.S. Environmental Protection Agency.

Baria, D.N. 1975. Survey of trace elements in North Dakota lignite and effluent streams from combustion and gasification facilities. Engineering Exp. Station, University of North Dakota. 64 pp.

Beckner, J.L. 1975. Trace element composition and disposal of gasifier ash. Seventh Synthetic Pipeline Gas Symposium, Chicago.

Beckner, J.L. 1976. Consider ash disposal (from gasification of coal). Hydrocarb. Process. 55(2):107-109.

Beychok, M.R. 1975. Perspective on coal conversion/letter. Environ. Sci. Technol. 9(5):396-397.

Coleman, W.M., P. Szabo, D.L. Wooten, H.C. Dorn, and L.T. Taylor. 1976. Minor and trace metal analysis of solvent refined coal by flameless atomic absorption. Fuel 56(2):195-198.

Davis, A., W. Spackman, and P.H. Given. 1976. The influence of the properties of coals on their conversion into clean fuels. Energy Sources 3(1):55-81.

Forney, A.J., W.P. Haynes, S.J. Gasior, G.E. Johnson, and J.P. Strakey, Jr. 1974. Analysis of tars, chars, gases, and water found in effluents from the synthane process. U.S. Bur. Mines Tech. Prog. Report 76. Pittsburgh Energy Research Center, U.S. Energy Research and Development Administration. 9 pp.

Forney, A.J., W.P. Haynes, S.J. Gasior, G.E. Johnson, and J.P. Strakey, Jr. 1975. Trace element and major component balances around the synthane PDU gasifier. U.S. Bur. Mines Tech. Prog. Report 75. PERC-TPR-75-1. Pittsburgh Energy Research Center, U.S. Energy Research and Development Administration.

Given, P.H., R.M. Miller, N. Suhr, and W. Spackman. 1975. Major, minor, and trace elements in the liquid product and solid residue from catalytic hydrogenation of coals. In Trace Elements in Fuel, S.P. Babu, ed. Advances in Chemistry Series No. 141, American Chemical Society, Washington, D.C., pp. 188-191.

Griffin, R.A., R.M. Schuller, J.J. Suloway, S.A. Russell, W.F. Childers, and N.F. Shimp. 1978. Solubility and toxicity of potential pollutants in solid coal wastes. In Proceedings of Third Symposium on the Environmental Aspects of Fuel Conversion Technology, Sept. 1977, Hollywood, Fla. F.A. Ayer and M.F. Massoglia, compilers. EPA-600/7-78-063. U.S. Environmental Protection Agency, pp. 506-518.

Institute of Gas Technology. 1976. Preparation of a coal conversion systems data book. FE 2286-4. U.S. Energy Research and Development Administration.

Jahnig, C.E. 1975a. Evaluation of pollution control in fossil fuel conversion processes. Gasification: Section 5. Bigas process. EPA-650/2-74-009-g. U.S. Environmental Protection Agency.

Jahnig, C.E. 1975b. Evaluation of pollution control in fossil fuel conversion processes. Gasification: Section 6. Hygas process. EPA-650/2-74-009-h. U.S. Environmental Protection Agency.

Jahnig, C.E. 1975c. Evaluation of pollution control in fossil fuel conversion processes. Gasification: Section 7. U-gas process. EPA-650/2-74-009-i. U.S. Environmental Protection Agency.

Jahnig, C.E., and R.R. Bertrand. 1976. Coal processing: Environmental aspects of coal gasification. Chem. Eng. Prog. 72(8):51-56.

Jahnig, C.E., and E.M. Magee. 1975. Evaluation of pollution control in fossil fuel conversion processes. Liquefaction: Section 2. SRC Process. EPA-650/2-74-009-f. U.S. Environmental Protection Agency.

Kalfadelis, C.D., and E.M. Magee. 1974. Evaluation of pollution control in fossil fuel conversion processes. Gasification: Section 1. Synthane process. EPA-650/2-74-009-b, U.S. Environmental Protection Agency, pp. 1-87.

Kalfadelis, C.D., E.M. Magee, G.E. Milliman, and T.D. Searl. 1975. Evaluation of pollution control in fossil fuel conversion processes. Analytical test plan. EPA-650/2-74-009-l. U.S. Environmental Protection Agency.

Kalika, P.W., P.T. Bartlett, R.E. Kenson, and J.E. Yocom. 1975. Measurement of fugitive emissions. Presented at 68th Annual Air Pollution Control Association Meeting, Boston, Mass.

Kenson, R.E., P.W. Kalika, and J.E. Yocom. 1975. Fugitive emissions from coal. Presented at National Coal Association/Bituminous Coal Research Coal Conf. and Expo. II, Louisville, Ky.

Kenson, R.E., N.E. Bowne, and W.A. Cote. 1976. The cost effectiveness of coke oven control technology--study of ambient air impact. Presented as APCA Western Penn. Section Conf., Pittsburgh, Pa.

Magee, E.M. 1976. Evaluation of pollution control in fossil fuel conversion processes. Final Report. EPA-600/2-76-101. U.S. Environmental Protection Agency.

Magee, E.M., H.J. Hall, and G.M. Varga, Jr. 1973. Potential pollutants in fossil fuels. EPA-R2-73-249. U.S. Environmental Protection Agency. 151 pp.

Magee, E.M., C.E. Jahnig, and H. Shaw. 1974. Evaluation of pollution control in fossil fuel conversion processes. Gasification: Section

1. Koppers-Totzek process. EPA-650/2-74-009-a. U.S. Environmental Protection Agency.

NRC Committee on Processing and Utilization of Fossil Fuels. 1977. Assessment of technology for the liquefaction of coal. Commission on Sociotechnical Systems. Department of Energy Report FE1216-3. Available from NTIS, Springfield, Va. 165 pp.

O'Hara, J.B., S.N. Rippee, B.I. Loran, and W.J. Mindheim. 1974. Environmental factors in coal liquefaction plant design. Presented at EPA Symposium on Environ. Aspects of Fuel Conv. Technol. (OCR R.D. Rep. 82. Interim Rep. No. 3, Contract 14-32-0001-1234). Ralph M. Parsons Co.

Parrish, R.W., and H.B. Neilson. 1974. Synthesis gas purification including removal of trace contaminants by the Benfield process. Indust. and Eng. Chem. Div., American Chemical Society.

Rubin, E.S., and F.C. McMichael. 1975. Impact of regulations on coal-conversion plants. Environ. Sci. Technol. 9(2):112-117.

Rubin, E.S., F.C. McMichael, and M.R. Beychok. 1975. Coal conversion retort/letter. Environ. Sci. Technol. 9(6):497.

Ruch, R.R., H.J. Gluskoter, G.B. Dreher, S.J. Russell, R.A. Cahill, J.K. Frost, J.K. Kuhn, J.D. Steele, J.F. Ashby, J. Thomas, and R. Malhotra. 1977. Determination of valuable metals in liquefaction process residues. Rep. No. 21. ERDA contract (E46-1)-8004. FE-8004-14, U.S. Department of Energy.

Sather, N.F., W.M. Swift, J.R. Jones, J.L. Beckner, J.H. Addington, and R.L. Wilburn. 1975. Potential trace element emissions from the gasification of Illinois coals. Final Rep. Prog. 11EE-80.026. Rep. No. 11EE-75-08. Illinois Institute for Environmental Quality. NTIS PB-241, 220/3ST. 18 pp.

Schultz, H., E.A. Hattman, and W.B. Booher. 1975. Trace elements in coal--what happens to them? Amer. Chem. Soc. Div. Environ. Chem., Prepr. 15(1):196-197.

Schultz, H., G.A. Gibbon, E.A. Hattman, H.B. Booher, and J.W. Adkins. 1977. The distribution of some trace elements in the 1/2 ton per day Synthoil process development unit. PERC/RI-77/2. U.S. Energy Research and Development Administration. 17 pp.

Sharkey, A.G., J.L. Shultz, C.E. Schmidt, and R.A. Friedel. 1975. Mass spectrometric analysis of stream from coal gasification and liquefaction processes. PERC/RI-75/5. U.S. Energy Research and Development Administration.

Shaw, H., and E.M. Magee. 1974. Evaluation of pollution control in fossil fuel conversion process. Gasification: Section 1. Lurgi process. EPA-650/2-74-009-R, U.S. Environmental Protection Agency, pp. 1-70.

Somerville, M.H., J.L. Elder, and R.G. Todd. 1977. Trace elements: analysis of their potential impact from a coal gasification facility. Eng. Exp. Sta. Bull. 77-05-EES-01. University of North Dakota.

U.S. Bureau of Reclamation. 1977. Draft environmental impact statement. ANG Coal Gasification Company, North Dakota project, Upper Missouri Region. U.S. Department of the Interior.

APPENDIXES

APPENDIX A 1975 Coal Flow Magnitudes[a]

Region	Mining Method ($\times 10^3$ tons)			Average Heat Value (Btu/lb)	Preparation[b] (percent processed)	Transportation[b] (percent of total shipments)			Electrical Utilities ($\times 10^3$ tons coal consumed)	Average Heat Value[c] (Btu/lb)	Coke Plants (coal capacity $\times 10^3$ tons)	Future Coal Conversion Plants	
	Underground	Strip	Strip Auger and Auger			Unit train	Barge	Truck				Liquefaction ($\times 10^6$ tons coal)	Gasification ($\times 10^6$ tons coal)
Northern Appalachia													
Maryland	104	2,466	0	11,344	79	24	3	31	3,372	12,479	4,710	0	1.1
Pennsylvania	44,631	39,105	402	11,825	89	22	21	27	36,140	11,960	13,875	0	NA
Ohio	15,455	24,908	6,407	10,909	77	36	13	29	46,167	11,262	4,009	306.2	2.1
Northern West Virginia	35,432	9,386	0	11,975	97	22	14	4	12,873	11,761	3,362	0	NA
TOTAL	95,622	75,765	6,809	11,614	88	26	17	22	98,552	11,625	2,596	306.2	3.2
Central Appalachia													
Eastern Kentucky	40,628	20,656	25,972	11,326	89	18	15	4	12,512	10,838	330	NA	0.8
Virginia	23,121	9,145	3,184	11,802	95	12	0	11	3,870	12,258	0	0	0
Southern West Virginia	52,905	7,560	0	11,975	97	22	14	4	12,873	11,761	3,362	0	NA
TOTAL	116,654	37,361	29,156	11,625	93	18	12	5	29,255	11,431	3,692	0	0.8
Central													
Indiana	188	24,935	0	10,811	99	39	14	19	25,980	10,811	9,538	0	0
Illinois	31,875	27,661	0	10,710	100	42	3	5	28,996	10,267	2,693	546	4.9
Western Kentucky	25,004	51,678	26,303	11,326	89	18	15	4	12,512	10,838	330	NA	0.8
TOTAL	57,067	104,274	26,303	11,062	94	28	11	7	67,488	10,582	12,561	546	5.7
Northwest (Powder River Basin)													
Montana	0	21,752	0	8,987	100	90	0	0	900	7,875	0	28	50.4
Wyoming	0	8,120	0	9,169	100	50	0	0	7,393	8,360	0	0	39
TOTAL	0	29,872	0	9,128	100	80	0	0	8,293	8,309	0	28	89.4
Southwest													
New Mexico	164	8,021	0	8,966	100	7	0	0	7,250	8,923	0	0	764
TOTAL	164	8,021	0	8,966	100	7	0	0	7,250	8,923	0	0	764

[a]Sources: U.S. Federal Power Commission (1976), U.S. Bureau of Mines (1976), Mining Information Services (1977), and Interagency Task Force on Synthetic Fuels (1975).

[b]1974 data.

[c]1973 data.

APPENDIX B Energy Content of Coal Flows in the United States in 1975, by Region and Process ($\times 10^{12}$ Btu)[a]

Region	Extraction[b] Under-ground	Surface	Auger	Preparation	Transportation Unit Train	Barge	Truck	Utilization Utilities	Coke	Liquefaction	Gasification
Northern Appalachia	2,000 (4,000)	2,000 (2,000)	200 (200)	4,000	1,000	700	900	2,000	700	7,000	70
Central Appalachia	3,000 (5,000)	900 (10,000)	700 (800)	4,000	700	500	200	700	100	0	20
Central	1,000 (2,000)	2,000 (3,000)	600 (700)	4,000	1,000	500	300	1,000	300	10,000	100
Northwest	0 (0)	500 (1,000)	0 (0)	500	400	0	0	100	0	500	1,000
Southwest	10 (30)	100 (200)	0 (0)	200	10	0	0	100	0	0	800

[a]Sources: Appendix A and Hittman Associates, Inc. (1977).

[b]For extraction, values are Btu's mined. Coal magnitude inputs for extraction are obtained by dividing outputs by extraction efficiencies and are shown in parentheses. Extraction efficiencies given by Hittman Associates (1977) are as follows:

	Underground	Surface	Auger
Northern Appalachia	0.57	0.81	NA, assume same as surface
Central Appalachia	NA, assume same as Northern Appalachia	0.81	NA, assume same as surface
Central	NA, assume same as Northern Appalachia	0.81	NA, assume same as surface
Northwest	—	0.98	
Southwest	NA, assume same as Northern Appalachia	0.79	

APPENDIX C

TRACE METALS MOBILIZED INTO AIR

The emissions magnitudes in Table C-1 are the products of energy inputs by process and region (Appendix B) and a set of specific emissions coefficients for air and water, by metal, process, and region. The energy-input figures are derived from government sources and are as reliable as such figures can be, but the emissions coefficients used (Hittman Associates, Inc., 1977) may be somewhat arbitrary.

Stated caveats in the Hittman report include restrictions in the availability and accessibility of technical data and derivation of coefficients from a limited number of cases. We have restricted our use of the coefficients to comparisons of order-of-magnitude ranges, for we believe the figures to be accurate only to that degree, at best. The coefficients cover effluents from active processes only (i.e., not acid mine drainage from inactive mines), and only the predominant modes of processes (e.g., unit train, barge, or truck for transportation) are considered.

The Hittman report provides emissions coefficients for both "uncontrolled" and "controlled" cases. We have used the controlled coefficients as the better indicator of present and future emissions, except for steam-electric plant water emissions, where no controlled emissions coefficients were given.

The regional energy inputs are aggregated to be consistent with the geographic composition of the regions as defined by Hittman, except that Eastern Kentucky is included in the Northern Appalachia region, because the coal seams are continuous with those of that region.

Readers who wish to quote emissions numbers or ratios from this report are urged to consult the Hittman report before doing so.

TABLE C.1 Amount of Trace Metals Mobilized into Air by Steam-Electric Plant Coal Combustion and by Coal Supply Processes[a] by U.S. Region and Metal, 1975 (U.S. tons)[b]

Metal	Appalachia		Central		Northwest		Southwest	
	Supply	Combustion	Supply	Combustion	Supply	Combustion	Supply	Combustion
Arsenic	0.9	1	0.5	0.7	0.03	0.007	0.01	0.006
Beryllium	0.2	4	0.07	3	0.003	0.03	0.0002	0.1
Cadmium	0.03	—	0.04	100	0.001	0.02	0.0003	0.002
Chromium	0.9	100	0.6	200	0.09	7	0.01	3
Copper	1	—	0.5	10	0.2	1	0.01	0.6
Iron	0.9	10,000	0.0003	10,000	0.00007	20	0.000003	200
Lead	0.6	80	0.8	400	0.06	1	0.008	5
Mercury	0.01	40	0.009	20	0.001	80	0.01	0.3
Nickel	1	500	0.2	400	0.05	2	0.01	2
Vanadium	2	100	1	70	0.3	300	0.02	3
Zinc	2	100	10	700	0.08	60	0.02	7

[a] Includes extraction (underground, strip, and auger) and transportation (unit train, barge, truck). Air emissions from preparation were reported as being zero.

[b] Source: After Hittman Associates, Inc. (1977), 12 unpublished appendixes (numbered 4-15), on file in GERHD office, NRC.

ABBREVIATIONS USED

Chemical elements and their symbols are shown on Figure 1 and Table 6.

Abbreviation	Expansion or equivalent
Å	angstrom
ABS	alkylbenzenesulfonate
ACS	American Chemical Society
$Al_6Si_2O_{13}$	mullite, an orthorhombic aluminum silicate
AMD	acid mine drainage
AsO_3	arsenate
ASTM	American Society for Testing Materials
BA	bottom ash
Btu	British thermal unit
°C	degrees, Celsius
CaO	calcium oxide
$CaSO_4$	calcium sulfate
Ca^{2+}	divalent calcium
cbar	centibar (10^{-2}) (one hundredth of a bar)
CFR	Code of Federal Regulations
CH_4	methane, a gaseous hydrocarbon
cm	centimeter (10^{-2})
CO	carbon monoxide
CO_χ	carbonaceous oxide
CWP	coal workers pneumoconiosis
DOE	Department of Energy
DTPA	diethylenepentamineacetic acid
EC_e	electrical conductivity of the extract
EPA	Environmental Protection Agency
ERDA	Energy Research and Development Administration (now Department of Energy)
FBB	fluidized-bed boiler
FeS_2	iron sulfide (pyrite)
FGD	flue-gas desulfurization
ft^3	cubic foot
FWPCA	Federal Water Pollution Control Administration
g	gram
G	gravity

GERHD	Subcommittee on the Geochemical Environment in Relation to Health and Disease
h	hour
ha	hectare (10,000 m^2 or 2.47 acres)
H_2	diatomic hydrogen
HEFF	Committee on Research Needs on the Health Effects of Fossil Fuel Combustion Products
H_2O	hydrogen oxide (water)
H_2S	hydrogen sulfide
H_2SO_4	sulfuric acid
IAEA	International Atomic Energy Agency
INIS	International Nuclear Information Service
ISO	International Standards Organization
kg	kilogram
km	kilometer
l	liter
LSD	least-significant difference
m^3	cubic meter
meq	milliequivalent
mg	milligram (10^{-3}) (one thousandth of a gram)
mi	mile
ml	milliliter (10^{-3}) (one thousandth of a liter)
mmho	millimho (10^{-3}) (one thousandth of a mho, the reciprocal of an ohm)
mrem	millirem (one thousandth of a roentgen equivalent man--an injury-dose measurement)
MW	megawatt (10^6) (one million watts)
MWe	megawatt of electricity
NAS	National Academy of Sciences
NBS	National Bureau of Standards
NH_4OA	ammonium acetate
NPDES	National Pollutant Discharge Elimination System
NRC	National Research Council
PC-fired	pulverized-coal-fired
pCi	picocurie (10^{-12}) (one millionth of a millionth)
PECH	Panel on Trace Element Geochemistry of Coal Resource Development Related to Health
pH	acid-base scale (0 to 14; 7 neutral)
PMF	progressive massive fibrosis
ppb	parts per billion
ppm	parts per million
R_{ij}	an enrichment ratio of trace elements in coal burning
s	second
SEM	scanning electron microscope
SeO_3	selenite ion
SO_2	sulfur dioxide
SO_χ	sulfurous oxides
SO_4^-S	sulfate sulfur expressed as elemental sulfur
SRC	solvent-refined coal
SRM	Standard Reference Material
TNT	trinitrotoluene, an explosive
USGS	U.S. Geological Survey
μm	micrometer (10^{-6}) (one millionth of a meter)

REFERENCES

Allaway, W.H. 1968. Agronomic controls over the environmental cycling of trace elements. Adv. Agron. 20:235-274.

Allegheny County Bureau of Air Pollution Control. 1978. Allegheny County Emission Inventory. Pittsburgh, Pa.

Amdur, M.O. 1976. Toxicological guidelines for research on sulfur oxides and particulates. In Proceedings of the Fourth Symposium on Statistics and the Environment, March 3-5, Washington, D.C.

American Chemical Society. 1977a. Preprints of papers presented at 174th National Meeting, Fuel Chemistry Division, Chicago, Aug. 29-Sept. 2, Vol. 22, No. 5. 183 pp.

American Chemical Society. 1977b. Application reviews. Anal. Chem. 49(5):1R-286R.

American Chemical Society. 1978. Fundamental reviews. Anal. Chem. 50:1R-384R.

American Society for Testing and Materials. 1973. Sampling, standards, and homogeneity. ASTM Spec. Tech. Pub. 540. Philadelphia, Pa. 136 pp.

Amick, R.S., K. Axetell, Jr., and D.M. Wells. 1974. Environmental Dust Emission Investory Techniques. Proceedings, 67th Annual Meeting of the Air Pollution Control Assn., Denver, Colo., June 9-13, 1974.

Anders, O.U. 1977. Representative sampling and proper use of reference materials. Anal. Chem. 49(1):33A-36A.

Anderson, W.L., and K.E. Smith. 1977. Dynamics of mercury at coal-fired power plant and adjacent coaling lake. Environ. Sci. Technol. 11:75-80.

Andren, A.W., B.G. Blaylock, E.A. Bondietti, C.W. Francis, S.G. Hildebrand, J.W. Huckabee, D.R. Jackson, S.E. Lindberg, F.H. Sweeton, R.I. Van Hook, and A.P. Watson. 1974. Ecology research. In Ecology and Analysis of Trace Contaminants. ORNL-NSF-EATC-11. National Science Foundation, Washington, D.C., pp. 61-104.

Attari, A. 1973. Fate of trace constituents of coal during gasification. EPA-650/2-73-004. U.S. Environmental Protection Agency. 31 pp.

Attari, A., M. Mensinger, and J.C. Pau. 1975. Fate of trace constituents in coal gasification. EPA-659/2-73-004. U.S. Environmental Protection Agency.

Averitt, P. 1975. Coal resources of the United States, January 1, 1974. U.S. Geol. Surv. Bull. 1412. 131 pp.

Averitt, P., I.A. Breger, V.E. Swanson, P. Zubovic, and H. Gluskoter. 1972. Minor elements in coal--a selected bibliography. U.S. Geol. Surv. Prof. Paper 800-D, pp. 169-171.

Averitt, P., J.R. Hatch. V.E. Swanson, I.A. Breger, S.L. Coleman, J. H. Medlin, P. Zubovic, and H.J. Gluskoter, compilers. 1976. Minor and trace elements in coal--a selected bibliography of reports in English. U.S. Geol. Surv. Open File Report 76-481. 16 pp.

Ayer, F.A. 1974. Symposium Proceedings: Environmental Aspects of Fuel Conversion Technology. EPA-650/2-74-118. U.S. Environmental Protection Agency. 365 pp.

Ayer, F.A. 1976. Symposium Proceedings: Environmental Aspects of Fuel Conversion Technology, II. EPA-600/2-76-149. U.S. Environmental Protection Agency. 403 pp.

Babu, S.P., ed. 1975. Trace Elements in Fuel. Advances in Chemistry Series No. 141. American Chemical Society, Washington, D.C. 216 pp.

Baria, D.N. 1975. Survey of trace elements in North Dakota lignite and effluent streams from combustion and gasification facilities. Engineering Exp. Station, University of North Dakota. 64 pp.

Baser, M.E., and S.C. Morris. 1977. Assessment of the potential role of trace metal health effects in limiting the use of coal-fired electric power. Informal Report, Brookhaven National Laboratory, Upton, N.Y.

Bayliss, R.J., and H.M. Whaite. 1966. A study of radium alpha activity of coal, coal ash, and particulate emission at a Sydney power station. Air Water Pollut. 10:813-819.

Beilke, S., and G. Gravenhorst. 1977. Heterogeneous SO_2 oxidation in the droplet phase. Prepared for presentation at the International Symposium on Sulfur in the Atmosphere, September 7-14, 1977, Dubrovnik, Yugoslavia.

Bern, J. 1976. Residues from power generation: processing, recycling, and disposal. In Land Application of Waste Materials. Soil Conservation Society of America, Ankeny, Iowa, pp. 226-248.

Berry, W.L., and A. Wallace. 1974. Trace elements in the environment--their role and potential toxicity as related to fossil fuels. A preliminary study. University of California, Los Angeles. Laboratory of Nuclear Medicine and Radiation Biology, Atomic Energy Commission Contract AT(04-1)GEN-12.

Bertine, K.K., and E.D. Goldberg. 1971. Fossil fuel combustion and the major sedimentary cycle. Science 173:233-235.

Bingham, F.T. 1973. Boron in cultivated soils and irrigation waters. In Trace Elements in the Environment. Advances in Chemistry Series No. 123. American Chemical Society, Washington, D.C., pp. 130-138.

Bird, B.R., W.E. Stewart, and E.N. Lightfood. 1960. Transport Phenomena, Chapter 16. John Wiley and Sons, Inc., New York, p. 495.

Block, C., and R. Dams. 1975. Inorganic composition of Belgian coals and coal ashes. Environ. Sci. Technol. 9:146-150.

Block, C., and R. Dams. 1976. Study of fly ash emission during combustion of coal. Environ. Sci. Technol. 10:1011-1017.

Bowen, H.J.M. 1966. Trace Elements in Biochemistry. Academic Press, New York.

Brackett, C.E. 1973. Production and utilization of fly ash in the United States. Proc. Third International Ash Utilization Symposium. Pittsburgh, Pa.

Bradford, G.R., F.L. Blair, and V. Hunsaker. 1971. Trace and major element contents of soil saturation extracts. Soil Sci. 112:225-230.

Bradford, G.R., A.L. Page, I.R. Straughan, and H.T. Phung. 1978. A study of the deposition of fly ash on desert soils and vegetation adjacent to a coal-fired generating station. In Environmental Chemistry and Cycling Process, D.C. Adriano and I.L. Brisbin, eds. DOE Symp., Ser. 45. CONF-760429. U.S. Department of Energy, Washington, D.C., pp. 383-393.

Brown, D. 1977. Handbook: Equipment for Reclaiming Strip Mined Land. U.S. Dept. of Agriculture, Forest Service, Equipment Development Center, 59801 (No. 7728 2503) Ft. Missoula, Missoula, Mont. February, p. 21.

Caldwell, R.D., R.F. Crosby, and M.P. Lockard. 1970. Radioactivity in coal mine drainage. In Environmental Surveillance in the Vicinity of Nuclear Facilities. W.C. Reing, ed. Charles C Thomas Publishers, Springfield, Ill., pp. 438-445.

Carlberg, J.R., J.V. Crable, L.P. Limtiaca, H.B. Norris, J.L. Holtz, P. Mauer, and F.R. Wolowicz. 1971. Total dust, coal, free silica and trace metal concentrations in bituminous coal miners' lungs. Am. Ind. Hyg. Assoc. J. 32:432.

Carter, M.D. 1976. National Coal Resource Data System of the U.S. Geological Survey. Computers in Geoscience 2:331-340.

Cassidy, S.M., ed. 1973. Elements of practical coal mining. Chapter 15, Modern Mining Methods--Surface. E.R. Phelps, ed. Soc. of Mining Engineers and Am. Inst. of Mining, Metallurgical and Petroleum Engs., Inc., New York.

Chang, A.C., L.J. Lund, A.L. Page, and J.E. Warneke. 1977. Physical properties of fly ash-amended soils. J. Environ. Qual. 6:267-270.

Coleman, W.M., P. Szabo, D.L. Wooten, H.C. Dorn, and L.T. Taylor. 1976. Minor and trace metal analysis of solvent refined coal by flameless atomic absorption. Fuel 56(2):195-198.

Collin, P.J. 1974. Some aspects of the chemistry of fly ash surfaces. Proceedings of Symposium on the Changing Technology of Electrostatic Precipitators, Adelaide. Institute of Fuel, Sydney, p. 16.

Cope, F. 1962. The development of a soil from an industrial waste ash. Int. Soc. Soil. Sci., Trans. Comm. IV, V (Palmerston North, New Zealand), pp. 859-863.

Council on Environmental Quality. 1973. Energy and the Environment. Electric Power. No. 4111-00019. U.S Government Printing Office, Washington, D.C., p. 40.

Crockett, A.B., and R.R. Kinnison. 1977. Mercury distribution in soil around a large coal-fired power plant. NTIS PB-269289. National Technical Information Service, Springfield, Va.

Davis, J.M.G., J. Ottery, and A. Le Roux. 1977. The effects of quartz and other non-coal dusts in coal workers' pneumoconiosis. Part II. Lung autopsy study. In Inhaled Particles IV, Part 2, W.H. Walton, ed. Pergamon Press, Oxford, pp. 691-702.

Davison, R.L., D.F.S. Natusch, J.R. Wallace, and C.A. Evans, Jr. 1974. Trace elements in fly ash, dependence of concentration on particle size. Environ. Sci. Technol. 8:1107-1113.

Dickson, E.M. 1972. Mercury and lead in the Greenland Ice Sheet: a reexamination of the data. Science 177:536.

Doran, J.W., and D.C. Martens. 1972. Molybdenum availability as influenced by application of fly ash to soil. J. Environ. Qual. 1:186-189.

Dreher, G.B. 1978. Determination of valuable metals in liquefaction process residues--Quarterly technical report for period 1-1-78 to 3-31-78. FE-8004-26. U.S. Department of Energy. 13 pp.

Drysdale, F.R., and C.E. Calef. 1977. *The Energetics of the United States of America: An Atlas*. BNL 50501-R. Brookhaven National Laboratory, Upton, N.Y.

Duce, R.A., G.L. Hoffman, and W.H. Zoller. 1975. Atmospheric trace metals at remote northern and southern hemisphere sites: pollution or natural? Science 187:59.

Ehrlich, S. 1976. Problems in coal blending. In *Coal Blending and Utilization Conf.*, Des Moines, Iowa, June 16-17, 1976.

Elseewi, A.A., F.T. Bingham, and A.L. Page. 1978a. Growth and mineral composition of lettuce and Swiss chard grown on fly ash amended soils. In *Environmental Chemistry and Cycling Process*, D.C. Adriano and I.L. Brisbin, eds. DOE Symp., Ser. 45. CONF-760429, U.S. Department of Energy, Washington, D.C., pp. 568-581.

Elseewi, A.A., F.T. Bingham, and A.L. Page. 1978b. Availability of sulfur in fly ash to plants. J. Environ. Qual. 7:69-73.

Elseewi, A.A., A.L. Page, and F.T. Bingham. 1978c. Availability of sulfur in sewage sludge to plants: A comparative study. J. Environ. Qual. 7:213-217.

Environmental Science and Technology. 1976. A southwest power plant saga, a special report. Environ. Sci. Technol. 10:532-537.

Fairman, R.P., R.J. O'Brien, S. Swecker, H.E. Amandus, and E.P. Shoub. 1977. Respiratory status of surface coal miners in the United States. Arch. Environ. Health 32(5):211.

Farmer, B.M., C.E. Styron, C.A. Phillips, B. Robinson, and W.H. Westendorf. 1977. The assessment of the radiological impact of western coal utilization, Phase I. USERDA Contract No. EY-76-C-04-0053. Monsanto Research Corporation, Mound Laboratory, Miamisburg, Ohio. 314 pp.

Filby, R.H., K.R. Shah, and C.A. Sautter. 1976. A study of trace element distribution in a coal conversion process using neutron activation analysis. In *Proc. 1976 Inst. Conf. Mod. Trends in Act. Anal.*, R. Henkelmann, ed. Munich, Sept. 13-17, 1976, pp. 664-674.

Fisher, G.L., D.P.Y. Chang, and M. Brummer. 1976. Fly ash collected from electrostatic precipitators. Microcrystalline structures and the mystery of the spheres. Science 192:553-555.

Flanagan, F.J. 1976. Descriptions and analyses of eight new USGS rock standards. U.S. Geol. Surv. Prof. Paper 840. U.S. Government Printing Office, Washington, D.C. 192 pp.

Fling, R.B. 1976. Disposal of flue gas cleaning wastes: EPA Shawnee field evaluation--initial report. EPA-600/2-76-070. U.S. Environmental Protection Agency, Washington, D.C.

Ford, C.T., R.R. Care, and R.E. Bosshart. 1976. Preliminary evaluation of the effect of coal cleaning on trace element removal. Trace Element Program, Report 3. Bituminous Coal Research, Inc., Monroeville, Pa. 115 pp.

Forney, A.J., W.P. Haynes, S.J. Gasior, G.E. Johnson, and J.P. Strakey, Jr. 1974. Analysis of tars, chars, gases, and water found in effluents from the synthane process. U.S. Bur. Mines Tech. Prog. Report 76. Pittsburgh Energy Research Center, U.S. Energy Research and Development Administration. 9 pp.

Forney, A.J., W.P. Haynes, S.J. Gasior, G.E. Johnson, and J.P. Strakey, Jr. 1975. Trace element and major component balances around the synthane PDU gasifier. U.S. Bur. Mines Tech. Prog. Report 75. PERC-TPR-75-1. Pittsburgh Energy Research Center, U.S. Energy Research and Development Administration.

Forney, A.J., W.P. Haynes, S.J. Gasior, R.M. Kornosky, C.E. Schmidt, and A.G. Sharkey. 1976. Trace elements and major component balances around the synthane PDU gasifier. In Symposium Proceedings: Environmental Aspects of Fuel Conversion Technology, II, F. Ayer, ed. EPA-600/2-76-149. U.S. Environmental Protection Agency, pp. 67-82.

Freedman, R.W., and A.G. Sharkey, Jr. 1972. Recent advances in analysis of respirable coal dust for free silica, trace elements and organic constituents. I. Coal-mine dust. Ann. N.Y. Acad. Sci. 200:7-16.

Furr, A.K., T.F. Parkinson, R.A. Hinrichs, D.R. Van Campen, C.A. Bache, W.H. Gutenmann, L.E. St. John, Jr., I.S. Pakkala, and D.J. Lisk. 1977. National survey of elements and radioactivity in fly ashes: absorption of elements by cabbage grown in fly ash-soil mixture. Environ. Sci. Technol. 11:1194-1201.

Furr, A.K., T.F. Parkinson, C.L. Heffron, J.T. Reid, W.M. Haschek, W.H. Gutenmann, C.A. Bache, L.E. St. John, Jr., and D.J. Lisk. 1978. Elemental contents of tissues and excreta of lambs, goats, and kids fed white sweet clover growing on fly ash. J. Agric. Food Chem. 26:847-851.

Gilluly, J., A.C. Waters, and A.O. Woodford. 1968. Principles of Geology. W.H. Freeman and Co., San Francisco.

Given, P.H. 1974. Problems in the chemistry and structure of coals as related to pollutants from conversion processes. In Symposium Proceedings: Environmental Aspects of Fuel Conversion Technology. EPA-650/2-74-118. U.S. Environmental Protection Agency, pp. 27-34.

Given, P.H., R.M. Miller, N. Suhr, and W. Spackman. 1975. Major, minor, and trace elements in the liquid product and solid residue from catalytic hydrogenation of coals. In Trace Elements in Fuel, S.P. Babu, ed. Advances in Chemistry Series No. 141, American Chemical Society, Washington, D.C., pp. 188-191.

Gleason, W.T. 1976. Western coal burning experience at Northern States Power Company. In Coal Blending and Utilization Conf., Des Moines, Iowa, June 16-17, 1976.

Gluskoter, H.J., R.A. Cahill, W.G. Miller, R.R. Ruch, and N.F. Shimp. 1976. An investigation of trace elements in coal. In Symposium Proceedings: Environmental Aspects of Fuel Conversion Technology, II, F. Ayer, ed. EPA-600/2-76-149. U.S. Environmental Protection Agency, pp. 38-46.

Gluskoter, H.J., R.R. Ruch, W.G. Miller, R.A. Cahill, G.B. Dreher, and J.K. Kuhn. 1977. Trace elements in coal: occurrence and distribution. Ill. State Geol. Surv. Circ. 499. Urbana, Ill. 154 pp.

Gluskoter, H.J., N.F. Shimp, and R.R. Ruch. 1980. Coal analyses, trace elements, and mineral matter, Chapter 7. In Chemistry of Coal

Utilization, 2nd Supplementary Volume, M.A. Elliott, ed. National Research Council, Wiley-Interscience, New York. In press.

Griffin, R.A., R.M. Schuller, J.J. Suloway, S.A. Russell, W.F. Childers, and N.F. Shimp. 1978. Solubility and toxicity of potential pollutants in solid coal wastes. In Proceedings of Third Symposium on the Environmental Aspects of Fuel Conversion Technology, Sept. 1977, Hollywood, Fla. F.A. Ayer and M.F. Massoglia, compilers. EPA-600/7-78-063. U.S. Environmental Protection Agency, pp. 506-518.

Guidoboni, R.J. 1973. Determination of trace elements in coal and coal ash by spark source mass spectrometry. Anal. Chem. 45:1275-1277.

Gutenmann, W.H., C.A. Bache, W.D. Youngs, and D.J. Lisk. 1976. Selenium in fly ash. Science 191:966-967.

Hattman, E.A., H. Schultz, and W.E. McKinstry. 1977. Solid and gaseous fuels. Anal. Chem. 49:176R-185R.

Heise, E.R., P.C. Major, M. Sharon Metnech, E.J. Parrish, Andrea L. Jordon, and W.K.C. Morgan. 1977. Predominance of histocompatibility antigens W18 and HL-A1 in miners resistant to complicated coal workers' pneumoconiosis. In Inhaled Particles IV, Part 2, W.H. Walton, ed. Pergamon Press, Oxford, pp. 495-507.

Heugi, T., and J. Jedwab. 1966. Distribution of radioactivity and optical phenomena in coal from Dorenaz (Canton of Valais, Switzerland). Schweiz. Mineral. Petrogr. Mitt. 46:377-387.

Higgins, B.P.J., S.C. Mohlegi, and R.L. Irvine. 1976. Lake treatment with fly ash, lime and gypsum. J. Water Pollut. Control Fed. 48:2153-2164.

Hittman Associates, Inc. 1977. Toxic trace pollutants coefficients for energy supply and conversion. Final Report HIT-705, ERDA Contract EX-77-C-03-1296. Columbia, Md.

Hodgson, D.R., and R. Holliday. 1966. The agronomic properties of pulverized fuel ash. Chem. Ind. 1966(20):785-790.

Holmes, J.A. 1918. The sampling of coal in the mine. U.S. Bur. Mines Tech. Paper 1. 22 pp.

Ignasiak, B.S., T.M. Ignasiak, and N. Berkowitz. 1975. Advances in coal analysis. Rev. Anal. Chem. 2:278-298.

Interagency Task Force on Synthetic Fuels. 1975. Synthetic Fuels Commercialization Program, Draft, Chapter IV.

Jahnig, C.E., and E.M. Magee. 1975. Evaluation of pollution control in fossil fuel conversion processes. Liquefaction: Section 2. SRC Process. EPA-650/2-74-009-f. U.S. Environmental Protection Agency.

Jurinak, J.J., W.J. Grenney, G.L. Woldridge, J.P. Riley, and R.J. Wagenet. 1977. A model of environmental transport of heavy metals originating from stack derived particulate emission in semi-arid regions. 77-RD-27, Research and Development. Southern California Edison Co., Rosemead, Calif.

Kaakinen, J.W., R.M. Jorden, M.H. Lawasani, and R.E. West. 1975. Trace element behavior in a coal-fired power plant. Environ. Sci. Technol. 9:862-869.

Kagey, B.T., B.G. Wixson, and N.L. Gale. 1978. Coal, water, trace elements and mortality. Proceedings of the Twelfth Annual Conference on Trace Substances in Environmental Health, June 5-8, 1978, D.D. Hemphill, ed. University of Missouri, Columbia.

Karr, C., Jr., ed. 1978. Analytical Methods for Coal and Coal Products. Academic Press, New York.

Kendrick, P., Jr. 1977. Acid mine drainage--an old problem with a new dimension. J. Water Pollut. Control Fed. 49:1576.

Kenson, R.E., N.E. Bowne, and W.A. Cote. 1976. The cost effectiveness of coke oven control technology--study of ambient air impact. Presented at APCA Western Penn. Section Conf., Pittsburgh, Pa.

Kirchhoff, W.H., ed. 1977. Methods and standards for environmental measurement. Nat. Bur. Standards Spec. Publ. 464. U.S. Government Printing Office, Washington, D.C. 691 pp.

Kirchner, H., E. Marz, and A. Schiffer. 1974. Radioactive emission from power stations fired with Rhineland soft coal. Braunkohle 26:340-345.

Klein, D.H., and P. Russell. 1973. Heavy metals: Fallout around a power plant. Environ. Sci. Technol. 7:357-358.

Klein, D.H., A.W. Andren, J.A. Carter, J.F. Emery, C. Feldman, W. Fulkerson, W.S. Lyon, J.C. Ogle, Y. Talmi, R.F. Van Hook, and N. Bolton. 1975. Pathways of thirty-seven trace elements through coal-fired power plants. Environ. Sci. Technol. 9:973.

Konieczynski, J. 1969. Problems of microelement analysis in coal and its by-products. Koks, Smola, Gaz 1969(14):13-18.

Koppenaal, D.W., and S.E. Manahan. 1976. Hazardous chemicals from coal conversion processes? Environ. Sci. Technol. 10:1104-1107.

Kothny, E.L., ed. 1973. Trace Elements in the Environment. Advances in Chemistry Series No. 123. American Chemical Society, Washington, D.C. 149 pp.

LaFleur, P.D., ed. 1976. Accuracy in trace analysis: Sampling, sample handling, analysis. Nat. Bur. Standards Spec. Publ. 422. U.S. Government Printing Office, Washington, D.C. 1304 pp.

Lakowicz, J.R., M. McNamara, and L. Stevenson. 1978. Particle-mediated membrane uptake of chemical carcinogens studied by fluorescence spectroscopy. Science 199:305.

Laskin, S., M. Kushner, A. Sellakumar, and G.V. Katz. 1975. Combined carcinogen--irritant animal inhalation studies. In Air Pollution and the Lung, E.F. Aharonson, A. Ben-David, and M.A. Klingberg, eds. Proceedings of the Twentieth Annual "OHOLO" Biological Conference, Ma'alst, Israel, March 16-19.

Lee, H., T.O. Peyton, R.V. Steele, and R.K. White. 1977. Potential radioactive pollutants resulting from expanded energy programs. Stanford Research Institute Contract No. 68-03-2375. Menlo Park, Calif.

Lee, R.E., Jr., and D.J. von Lehmden. 1973. Trace metal pollution in the environment. J. Air Pollut. Control Assoc. 23:853-857.

Lee, R.E., H.L. Crist, A.E. Riley, and K.E. Macleod. 1975. Concentration and size of trace metal emissions from a power plant, a steel plant, and a cotton gin. Environ. Sci. Technol. 9:643-647.

Linton, R.W., H. Loh, D.F.S. Natusch, C.A. Evans, Jr., and P. Williams. 1976. Surface predominance of trace elements in airborne particles. Science 191:852-854.

Lisk, D.J. 1972. Trace metals in soils, plants, and animals. Adv. Agron. 24:267-325.

Lyon, W.S. 1977. Trace element measurements at the coal-fired steam plant. CRC Press, Cleveland, Ohio.

Magee, E.M., H.J. Hall, and G.M. Varga, Jr. 1973. Potential pollutants in fossil fuels. EPA-R2-73-249. U.S. Environmental Protection Agency. 151 pp.

Maienthal, E.J., and D.A. Becker. 1976. A survey of current literature on sampling, sample handling, and long term storage for environmental materials. Nat. Bur. Standards. Tech. Note 929. 34 pp.

Marquardt, W., R. Hoehle, and U. Schuh. 1970. Radioactive emissions by coal-fed power stations. Z. Gesamte Hyg. Ihre Grenzgeb. 16:188-191.

Martens, D.C. 1971. Availability of plant nutrients in fly ash. Compost Sci. 12:15-19.

Martens, D.C., M.G. Schnappinger, Jr., and L.W. Zelanzy. 1970. The plant availability of potassium in fly ash. Soil Sci. Soc. Am. Proc. 34:453-456.

McNay, L.M. 1971. Coal refuse fires, an environmental hazard. Informational Circular 9515. U.S. Bureau of Mines, Washington, D.C.

Meinke, W.W., and J.K. Taylor, eds. 1972. Analytical chemistry: Key to progress on national problems. Nat. Bur. Standards Spec. Publ. 351. U.S. Government Printing Office, Washington, D.C. 470 pp.

Mining Information Services. 1977. Keystone Coal Industry Manual. McGraw-Hill, New York. 1184 pp.

Morgan, W.K.C., R. Reger, D.B. Burgess, and E. Shoub. 1972. A comparison of the prevalence of coal workers' pneumoconiosis and respiratory impairment in Pennsylvania bituminous and anthracite miners. In Coal Workers Pneumoconiosis. I.J. Selikoff, M.M. Key, and D.H.K. Lee, eds. Ann. N.Y. Acad. Sci. 200:252-259.

Mulford, F.R., and D.C. Martens. 1971. Response of alfalfa to boron in fly ash. Soil Sci. Soc. Am. Proc. 35:296-300.

Naeye, R.L. 1972. Types of fibrosis in coal workers' pneumoconiosis. In Coal Workers' Pneumoconiosis, I.J. Selikoff, M.M. Key, and D.H.K. Lee, eds. Ann. N.Y. Acad. Sci. 220:381-400.

National Ash Association. 1975. Ash collections soar to 59.5 million tons in 1974. Ash at Work. VII:3. National Ash Association, Washington, D.C.

National Bureau of Standards. 1975. Catalog of NBS standard reference materials, 1975-1976 ed. NBS Special Publ. 260. 88 pp.

Natusch, D.F.S., and J.R. Wallace. 1974. Urban aerosol toxicity: the influence of particle size. Science 186:695-699.

Natusch, D.F.S., J.R. Wallace, and C.A. Evans, Jr. 1974. Toxic trace elements: preferential concentration in respirable particles. Science 183:202.

Natusch, D.F.S., C.F. Bauer, H. Matusiewicz, C.A. Evans, J. Baker, A. Loh, R.W. Linton, and P.K. Hopke. 1975. Characterization of trace elements in fly ash. Proceedings, International Conference on Heavy Metals in the Environment, Vol. II, Part 2, Toronto, Ontario, Canada, pp. 553-575.

Nordberg, G.F., ed. 1974. Effects and dose-response relationships of toxic metals. Proceedings from an International Meeting Organized by the Subcommittee on the Toxicology of Metals of the Permanent Commission and International Association on Occupational Health, Tokyo, November 18-23.

NRC Committee on Accessory Elements. 1979. Redistribution of Accessory Elements in Mining and Mineral Processing, Vol. I. Board on Mineral and Energy Resources, NRC Commission on Natural Resources. National Academy of Sciences, Washington, D.C.

NRC Study Committee to Assess the Feasibility of Returning Underground Coal Mine Wastes to the Mined-Out Areas. 1975. Underground Disposal of Coal Mine Wastes. NRC Environmental Studies Board, Board on Energy Studies. National Academy of Sciences, Washington, D.C. 172 pp.

Office of the Federal Register. 1978. Code of Federal Regulations, 1978 ed. Title 30. Mineral Resources. Chapter II, U.S. Geological Survey. National Archives and Records Service, General Services Administration. U.S. Government Printing Office, Washington, D.C. 982 pp.

Office of the Federal Register. 1979. Code of Federal Regulations, 1979 ed. Title 30. Mineral Resources. Chapter VII, Office of Surface Mining. National Archives and Records Service, General Services Administration. U.S. Government Printing Office, Washington, D.C. 1342 pp.

Ohio Basin Regional Federal Water Pollution Control Administration. 1969. Stream pollution by coal mine drainage in Appalachia. U.S. Dept. of the Interior, Cincinnati, Ohio.

Ondov, J.M., W.H. Zoller, I. Olmez, N.K. Aras, G.E. Gordon, L.A. Rancitelli, K.H. Abel, R.H. Filby, K.R. Shah, and R.C. Ragaini. 1975. Elemental concentrations in the National Bureau of Standards environmental coal and fly ash standard reference materials. Anal. Chem. 47:1102-1109.

Ondov, J.M., R.C. Regaini, R.E. Heft, G.L. Fisher, D. Silberman, and B.A. Prentice. 1976. Interlaboratory comparison of neutron activation and atomic absorption analyses of size-classified stack fly ash. 8th Materials Research Symposium Methods and Standards for Environmental Measurement, Sept. 20-24, 1976, Gaithersburg, Md.

Page, A.L., F.T. Bingham, L.J. Lund, G.R. Bradford, and A.A. Elseewi. 1977. Consequence of trace element enrichment of soils and vegetation from the combustion of fuel used in power generation. Biannual Report, Southern California Edison Research and Development Series 77-RD-29, Rosemead, Calif.

Page, A.L., A.A. Elseewi, and I. Straughan. 1979. Physical and chemical properties of fly ash from coal fired power plants with reference to environmental impacts. Residue Reviews 71. Springer-Verlag, New York.

Paulson, C.A.J., and A.R. Ramsden. 1970. Some microscopic features of fly ash particles and their significance in relation to electrostatic precipitation. Atmos. Environ. 4:175-185.

Peyton, T.O. 1976. Health implications of coal bearing uranium: a preliminary evaluation. TM-13100-2. Greenfield, Attaway & Tyler, Inc., San Rafael, Calif. 21 pp.

Phung, H.T., L.J. Lund, and A.L. Page. 1978. Potential use of fly ash as a liming material. In Environmental Chemistry and Cycling Process, D.C. Adriano and I.L. Brisbin, eds. DOE Symp. Ser. 45. CONF-760429, U.S. Department of Energy, Washington, D.C., pp. 504-515.

Plank, C.O., and D.C. Martens. 1974. Boron availability as influenced by application of fly ash to soil. Soil Sci. Soc. Am. Proc. 38:974-977.

Plank, C.O., D.C. Martens, and D.L. Hallock. 1975. Effect of soil application of fly ash on chemical composition and yield of corn

(Zea mays L.) and on chemical composition of displaced soil solutions. Plant and Soil 42:465-476.

Pollock, E.N. 1975. Trace impurities in coal by wet chemical methods. In Trace Elements in Fuel, S.P. Babu, ed. Advances in Chemistry Series No. 141, American Chemical Society, Washington, D.C., pp. 23-34.

Radian Corporation. 1975. The environmental effects of trace elements in the pond disposal of ash and flue gas desulfurization sludge. Southern California Edison Co., Rosemead, Calif.

Redmond, C.K., A. Ciocco, J.W. Lloyd, and H.W. Rush. 1972. Long-term mortality study of steel-workers, VI--mortality from malignant neoplasms among coke oven workers. J. Occup. Med. 17:621.

Rees, W.I., and G.H. Sidrak. 1956. Plant nutrition on fly ash. Plant and Soil 8:141-159.

Richards, L.A., ed. 1954. Diagnosis and Improvement of Saline and Alkali Soils. USDA Handbook No. 60. U.S. Department of Agriculture.

Rock, R.L., G. Svilar, R.T. Beckman, and D.D. Rapp. 1975. Evaluation of radioactive aerosols in United States underground coal mines. Mining Enforcement and Safety Administration Informational Report 1025. U.S. Department of the Interior. Washington, D.C.

Ross, R.H. 1977. Trace elements fate in soil. In Environmental Health and Control Aspects of Coal Conversion: An Information Overview. H.M. Braunstein, E.D. Copenhaver, and H.A. Pfunderer, eds. ORNL/ElS-95, June 24 and 25. Oak Ridge National Laboratory.

Ruch, R.R., H.J. Gluskoter, and N.F. Shimp. 1974. Occurrence and distribution of potentially volatile trace elements in coal: a final report. Ill. State Geol. Surv. Environ. Geol. Notes 72. 96 pp.

Safonov, I.S. 1974. Natural radioactivity of ash from heating and electric-power plants. Elektr. Stn. 3:52-53.

Sather, N.F., W.M. Swift, J.R. Jones, J.L. Beckner, J.H. Addington, and R.L. Wilburn. 1975. Potential trace element emissions from the gasification of Illinois coals. Final Rep. Prog. 1lEE-80.026. Rep. No. 1lEE-75-08. Illinois Institute for Environmental Quality. NTIS PB-241, 220/3ST. 18 pp.

Schnappinger, M.G., Jr., D.C. Martens, and C.O. Plank. 1975. Zinc availability as influenced by application of fly ash to soil. Environ. Sci. Technol. 9:258-261.

Schroeder, H.A. 1973. Recondite toxicity of trace elements. Essays Toxicol. 4:107-199.

Schultz, H., E.A. Hattman, W.B. Booher. 1975. The fate of some trace elements during coal pretreatment and combustion. In Trace Elements in Fuel, S.P. Babu, ed. Advances in Chemistry Series No. 141, American Chemical Society, Washington, D.C., pp. 139-153.

Schultz, H., G.A. Gibbon, E.A. Hattman, H.B. Booher, and J.W. Adkins. 1977. The distribution of some trace elements in the 1/2 ton per day Synthoil process development unit. PERC/RI-77/2. U.S. Energy Research and Development Administration. 17 pp.

Schwitzgebel, K., F.B. Meserol, R.G. Oldham, R.A. Magee, F.G. Mesich, and T.L. Thoem. 1975. Trace element discharge from coal-fired power plants. Proceedings, International Conference on Heavy Metals in the Environment, Vol. II, Part 2, Toronto, Ontario, Canada, pp. 543-551.

Sharkey, A.G., J.L. Shultz, C.E. Schmidt, and R.A. Friedel. 1975. Mass spectrometric analysis of stream from coal gasification and liquefaction processes. PERC/RI-75/5. U.S. Energy Research and Development Administration.

Sklyarenko, I.P., P.A. Enyakova, F.E. Krigman, M.D. Krivitski, A.P. Kolesnik, and A.K. Tsapuk. 1964. Investigations on the radioactivity of rocks in connection with their possible influence on the development of pneumoconiosis. Gig. Tr. Prof. Zabol. 8:13-19 (in Russian).

Somerville, M.H., J.L. Elder, and R.G. Todd. 1977. Trace elements: analysis of their potential impact from a coal gasification facility. Eng. Exp. Stn. Bull. 77-05-EES-01. University of North Dakota.

Stanford Research Institute. 1977. Potential radioactive pollutants resulting from expanded energy programs. USEPA Contract No. 68-03-2375. SRI Project EGU-4869. CRESS Rep. No. 6. Menlo Park, Calif. 130 pp.

Stoewsand, G.S., W.H. Gutenmann, and D.J. Lisk. 1978. Wheat grown on fly ash: High selenium uptake and response when fed to Japanese quail. J. Agric. Food Chem. 26:757-759.

Straughan, I., A.A. Elseewi, and A.L. Page. 1978. Mobilization of selected trace elements in residues from coal combustion with special reference to fly ash. In Proceedings of Twelfth Annual Conference on Trace Substances in Environmental Health, June 5-8, 1978, D.D. Hemphill, ed. University of Missouri, Columbia, pp. 389-402.

Swaine, D.J. 1977. Trace elements in coal. In Proceedings of the Eleventh Annual Conference on Trace Substances in Environmental Health, June 7-9, 1977, D.D. Hemphill, ed. University of Missouri, Columbia, pp. 107-116.

Swanson, V.E. 1972. Composition and trace element content of coal and power plant ash. Southwest Energy Study, Part 2, Appendix J. U.S. Geol. Surv. Open File Rep. 61 pp.

Swanson, V.E., and C. Huffman, Jr. 1976. Guidelines for sample collecting and analytical methods used in the U.S. Geological Survey for determining chemical composition of coal. U.S. Geol. Surv. Circ. 735. 11 pp.

Swanson, V.E., H.L. Cannon, P.H. Given, J.R. Jones, J.W. Kaakinen, H.E. Malde, S.B. McCaleb, J.Y. Parce, and T.L. Thoem. 1974. Environmental impact of conversion from gas or oil to coal for fuel. Geological Society of America. Geologist 9(4):1-8.

Swanson, V.E., J.H. Medlin, J.R. Hatch, S.L. Coleman, G.H. Wood, S.D. Woodruff, and R.T. Hildebrand. 1976. Collection, chemical analysis, and evaluation of coal samples in 1975. U.S. Geol. Surv. Open File Rep. 76-468. 503 pp.

Sweet, D.V., W.E. Crouse, J.V. Crable, J.R. Carlberg, and W.S. Lainhart. 1974. The relationship of total dust, free silica, and trace metal concentrations to the occupational respiratory disease of bituminous coal miners. Am. Ind. Hyg. Assoc. J. 35:479.

Sweet, D.V., W.E. Crouse, and J.V. Crable. 1978. Chemical and statistical studies of contaminants in urban lungs. Am. Ind. Hyg. Assoc. J. 39:515-526.

Szabo, M.F. 1978. Environmental assessment of coal transport. EPA 600/7-78-081. U.S. Environmental Protection Agency, Cincinnati, Ohio, pp. 75-78.

Taylor, S.R. 1964. Abundance of chemical elements in the continental crust: a new table. Geochim. Cosmochim. Acta 28:1273-1285.

Tenney, M.W., and W.F. Echelberger, Jr. 1970. Fly ash utilization in the treatment of polluted waters. U.S. Bur. Mines Info. Circ. 8488, pp. 237-268.

Thomson, J.M. 1963. Mortality threshold of fish in fly ash suspension. Aust. J. Sci. 25:414-415.

U.S. Bureau of Mines. 1976. Minerals Yearbook, 1974. Vol. I, Metals, Minerals, and Fuels. U.S. Government Printing Office, Washington, D.C. 1451 pp.

U.S. Bureau of Mines. 1977. Minerals Yearbook, 1975. Vol. I, Metals, Minerals, and Fuels. U.S. Government Printing Office, Washington, D.C. 1550 pp.

U.S. Bureau of Reclamation. 1977. Draft environmental impact statement. ANG Coal Gasification Company, North Dakota project, Upper Missouri Region. U.S. Department of the Interior.

U.S. Congress. 1972. Federal Water Pollution Control Act, Amendments of 1972. Public Law 92-500. National Pollution Discharge Elimination System. U.S. Government Printing Office, Washington, D.C.

U.S. Congress. 1976. Federal Coal Leasing Amendments of 1975. Public Law 94-377, August 4. U.S. Government Printing Office, Washington, D.C., pp. 565-568.

U.S. Congress. 1977. Surface Mining Control and Reclamation Act of 1977. Public Law 95-87, August 3. U.S. Government Printing Office, Washington, D.C.

U.S. Department of Health, Education, and Welfare. 1969. Air quality criteria for particulate matter. National Air Pollution Control Administration Publication No. AP-49, p. 115.

U.S. Energy Research and Development Administration. 1976. Sampling, strategy and characterization of potential emissions from synfuel production. CONF-760602. 146 pp.

U.S. Environmental Protection Agency. 1973. Water quality criteria 1972. EPA-R3-73-033. Washington, D.C.

U.S. Environmental Protection Agency. 1974. Methods for chemical analysis of water and waste. EPA-625/6-74-003. 298 pp.

U.S. Environmental Protection Agency. 1975. Coal Fired Power Plant Trace Element Study, A Three Station Comparison, Vols. I and II. Reports prepared by Radian Corporation. EPA Rocky Mountain Prairie Region VIII.

U.S. Federal Power Commission. 1976. Steam-electric air and water quality control data for the year ending December 31, 1973--Based on FPC Form No. 67; Summary Report. U.S. Government Printing Office, Washington, D.C.

U.S. Public Health Service. 1962. Public Health Service Drinking Water Standards. U.S. Government Printing Office, Washington, D.C.

Van Hook, R.I., and W.D. Shults, eds. 1976. Effects of trace contaminants from coal combustion. Proceedings of a Workshop, Aug. 2-6, 1976, Knoxville, Tenn. ERDA 77-64. National Technical Information Service.

Vasil'eva, I.V., and V.A. Vekhov. 1971. Behavior of beryllium during pyrolysis of coals. Khim. Tverd. Topl. 3:131-132.

Vaughan, B.E., K.H. Abel, D.A. Cataldo, J.M. Hales, C.E. Hane, L.A. Rancitelli, R.C. Routson, R.E. Wildung, and E.G. Wolf. 1975. Review of potential impact on health and environmental quality from metals entering the environment as a result of coal utilization. Pacific Northwest Laboratories, Battelle Memorial Institute, Richland, Wash.

Vogel, G.J., I. Johnson, and P. Cunningham. 1976. Development program on pressurized fluidized-bed combustion. Quarterly report for January-March, 1976. ANL/ES-CEN-1015. Argonne National Laboratory. 125 pp.

Wagner, J.C. 1972. Etiological factors in complicated coal workers' pneumoconiosis. In Coal Workers' Pneumoconiosis, I.J. Selikoff, M.M. Key, and D.H.K. Lee, eds. Ann. N.Y. Acad. Sci. 200:401-404.

Walton, W.H., J. Dodgson, G.G. Hadden, and M. Jacobsen. 1977. The effect of quartz and other non-coal dusts in coal workers' pneumoconiosis. Part I, Epidemiological studies. In Inhaled Particles IV, Part 2, W.H. Walton, ed. Pergamon Press, Oxford, pp. 669-690.

Waters, M.D. 1977. Toxicology of vanadium. In Toxicology of Trace Elements, R.A. Goyer and M.A. Mehlman, eds. John Wiley and Sons, New York.

Weaver, C. 1967. The determination of zinc in coal. Fuel 46:407-414.

Weiss, H.V., M. Koide, and E.D. Goldberg. 1971. Selenium and sulfur in a Greenland ice sheet: relation to fossil fuel combustion. Science 172:261.

Wewerka, E.M., J.M. Williams, N.E. Vanderburgh. 1976a. Disposal of coal preparation wastes: environmental considerations. National Conference on Energy and the Environment, Cincinnati, Ohio, Oct. 1976.

Wewerka, E.M., J.M. Williams, P.L. Wanek, and J.D. Olsen. 1976b. Environmental contamination from trace elements in coal preparation wastes: a literature review and assessment. EPA-600/7-76-007. U.S. Environmental Protection Agency and U.S. Energy Research and Development Administration, Washington, D.C.

Zoller, W.H., E.S. Gladney, and R.A. Dace. 1974. Atmospheric concentrations and sources of trace metals at the South Pole. Science 183:198.

Zoller, W.H., E.S. Gladney, G.E. Gordon, and J.J. Bors. 1975. Emissions of trace elements from coal-fired power plants. In Proceedings of the Eighth Annual Conference on Trace Substances in Environmental Health, June 11-13, 1974, D.D. Hemphill, ed. University of Missouri, Columbia, pp. 167-172.

Zubovic, P. 1976. Geochemistry of trace elements in coal. In Symposium Proceedings: Environmental Aspects of Fuel Conversion Technology, II. EPA-600/2-76-149. U.S. Environmental Protection Agency, pp. 47-63.

INDEX

The underscored numbers indicate the start of an extensive discussion of the topic.